*Quantitative
Techniques
for the*

# ANALYSIS
# OF
# SEDIMENTS

AN INTERNATIONAL SYMPOSIUM

# COMPUTERS and GEOLOGY

*a series edited by* Daniel F. Merriam

---

*1976-Quantitative Techniques for the Analysis of Sediments*

Computers and Geosciences - An International journal devoted to the rapid publication of computer programs in widely used languages and their applications

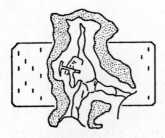

*Quantitative Techniques for the*
# ANALYSIS OF SEDIMENTS

AN INTERNATIONAL SYMPOSIUM

PROCEEDINGS OF AN INTERNATIONAL SYMPOSIUM HELD
AT THE IX INTERNATIONAL SEDIMENTOLOGICAL CONGRESS
IN NICE, FRANCE, ON 8 JULY 1975. THE MEETING WAS COSPONSORED
BY THE INTERNATIONAL ASSOCIATION FOR MATHEMATICAL GEOLOGY

*edited by*
DANIEL F. MERRIAM
*Jessie Page Heroy Professor of Geology and Chairman,
Department of Geology, Syracuse University, Syracuse, New York*

PERGAMON PRESS
OXFORD · NEW YORK · TORONTO · SYDNEY · PARIS · FRANKFURT

| | |
|---|---|
| U.K. | Pergamon Press Ltd., Headington Hill Hall, Oxford OX3 0BW, England |
| U.S.A. | Pergamon Press Inc., Maxwell House, Fairview Park, Elmsford, New York 10523, U.S.A. |
| CANADA | Pergamon of Canada Ltd., P.O. Box 9600, Don Mills M3C 2T9, Ontario, Canada |
| AUSTRALIA | Pergamon Press (Aust.) Pty. Ltd., 19a Boundary Street, Rushcutters Bay, N.S.W. 2011, Australia |
| FRANCE | Pergamon Press SARL, 24 rue des Ecoles, 75240 Paris, Cedex 05, France |
| WEST GERMANY | Pergamon Press GmbH, 6242 Kronberg-Taunus, Pferdstrasse 1, Frankfurt-am-Main, West Germany |

Copyright © 1976 Pergamon Press Ltd

*All Rights Reserved. No part of this publication may be reproduced, stored in a retrieval system or transmitted in any form or by any means: electronic, electrostatic, magnetic tape, mechanical, photocopying, recording or otherwise, without permission in writing from the publishers*

First edition 1976

**Library of Congress Cataloging in Publication Data**

Main entry under title:

Quantitative techniques for the analysis of sediments

(Computers & geology)
Proceedings of the 2d of a series of meetings; proceedings of the 1st are entered under Mathematical models of sedimentary processes
1. Marine sediments--Statistical methods--Congresses  2. Marine sediments--Mathematical models--Congresses  I. Merriam, Daniel Francis, 1927-
II. International Sedimentological Congress, 9th, Nice, 1975.
III. International Association for Mathematical Geology.
GC377.Q3 1976    551.4'6083    75-44011
ISBN 0-08-020613-1

*In order to make this volume available as economically and rapidly as possible the author's typescript has been reproduced in its original form. This method unfortunately has its typographical limitations but it is hoped that they in no way distract the reader.*

*Printed in Great Britain by A. Wheaton & Co. Exeter*

*dedicated to*
*those unselfish sedimentologists who*
*diligently*
*dedicatedly, and*
*deftly*
*studied the beach phenomena at*
*Nice and St. Tropez*
*in spite of*
*continual and*
*constant diversions*

# CONTENTS

List of Contributors ix

Preface xi

An analysis and management system suitable for sedimentological data, by J.M. Cubitt .......... 1

Trend analysis of sedimentary thickness data: the Pennsylvanian of Kansas, an example, by K.P. Thrivikramaji and D.F. Merriam .......... 11

Numerical classification of multivariate petrographic presence-absence data by association analysis in the study of the Miocene Ziqlag reef complex of Israel, by B. Buchbinder and D. Gill .................................. 23

An assessment of some quantitative methods of comparing lithological succession data, by W.A. Read ................................. 33

Statistical recognition of terrestrial and marine sediments in the Lower Cretaceous of Portugal, by R.A. Reyment, P.Y. Berthou, and B.Å. Moberg ..... 53

Classification of Minnesota lakes by Q- and R-mode factor analysis of sediment mineralogy and geochemistry, by W.E. Dean and E. Gorham ........ 61

Sedimentary environmental analysis of Long Island Sound, USA with multivariate statistics, by P.H. Feldhausen and S.A. Ali .......... 73

Simulation technique of matching and its stability, by D. Marsal ..................... 99

Mathematical modeling of sediment accumulation in prograding deltaic systems, by D.H. Horowitz ..... 105

A sedimentological pattern recognition problem, by M.W. Clark and I. Clark .............. 121

Multidimensional scaling of sedimentary rock descriptors, by J.H. Doveton ................. 143

The identification of discontinuities from areally distributed data, by S. Henley ........... 157

The display of three-factor models, by W.E. Stephens .. 169

Index ................................... 173

# LIST OF CONTRIBUTORS

Syed A. Ali, Department of Geology, Marine Science Research Center, SUNY Stony Brook, Stony Brook, New York 11794; present address, Department of Geology and Geophysics, Woods Hole Oceanographic Institute, Woods Hole, Massachusetts 02543, USA

P.Y. Berthou, Laboratoire de Geologie des Bassins Sedimentaires, Universite Paris VI, 4, Place Jussieu, 75230 Paris Cedex 05, France.

Binyamin Buchbinder, Geological Survey of Israel, 30 Malkhei Israel Street, Jerusalem 95 501, Israel

Isobel Clark, Department of Mining and Mineral Technology, Imperial College, University of London, London, England, UK

Malcolm W. Clark, Department of Geography, London School of Economics, University of London, London, England, UK

John M. Cubitt, Department of Geology, Syracuse University, Syracuse, New York 13210, USA

Walter E. Dean, Department of Geology, Syracuse University, Syracuse, New York 13210, USA

John H. Doveton, Kansas Geological Survey, University of Kansas, 1930 Avenue "A", Campus West, Lawrence, Kansas 66044, USA

Peter H. Feldhausen, Dames & Moore, P.O. Box 1633, Tehran, Iran

Dan Gill, Geological Survey of Israel, 30 Malkhei Israel Street, Jerusalem 95 501, Israel

Eville Gorham, Department of Ecology, University of Minnesota, St. Paul, Minnesota 55101, USA

Stephen Henley, Institute of Geological Sciences, Edinburgh, Scotland, UK

Daniel H. Horowitz, Exxon Production Research Company, P.O. Box 2189, Houston, Texas 77001, USA

Dietrich Marsal, Gewerkschaften Brigitta und Elwerath Betriebsfuhrungsgesellschaft mbH, 3 Hannover, Hinderburgstr. 28, Western-Germany

D.F. Merriam, Department of Geology, Syracuse University, Syracuse, New York 13210, USA

B.Å. Moberg, Paleontologiska Institutionen, Uppsala Universitet, Box 558, S75122, Uppsala, Sweden

William A. Read, Institute of Geological Sciences, London, UK

Richard A. Reyment, Paleontologiska Institutionen, Uppsala Universitet, Box 558, S75122, Uppsala, Sweden

W.E. Stephens, Department of Geology, University of St. Andrews, St. Andrews, Fife, Scotland, UK

K.P. Thrivikramaji, Department of Geology, Syracuse University, Syracuse, New York 13210, USA

# PREFACE

The IAMG session at SEDIM NICE (the 9th) was the second one held in conjunction with an IAS Congress - the first was at Heidelberg (the 8th) in 1971. Because sedimentologists were among the first in the earth sciences to use numbers in their analyses, it is appropriate that periodically reports be given on the use of statistics, mathematics, and computers in their work. It is equally important that they should be aware of advances outside the field and learn of possible applications in the development of others. It was with these objectives in mind that the subject "Quantitative Techniques for the Analysis of Sediments" was proposed.

To give a broad spectrum of involvement much effort was devoted to covering as many of the different techniques applicable as possible. But similar to many other endeavors, the ideal was too ambitious and in the final offering only a partial covering could be obtained. This sampling, however, should give the practitioner and others a good indication as to the level of sophistication in the subject as of the mid-70's.

In the 20 years of serious use of "numberology" in sedimentology we have seen a progression of methods tested and routinely used - from statistical examination of grains and the property of their distribution, through regional trend in sediment properties and studies on the relationship between sediment parameters, to simulation of sedimentary processes. This volume contains examples from each area and the reader thus can project future uses.

Some of the papers in this volume were not presented, some of the oral presentations are not recorded here, and some have been modified. Limited time of presentation at the meeting precluded no discussion. Unfortunately much of the value of the session was lost because of this restriction but hopefully the exchange can be in the written form, albeit much later.

Thus, the contents of this volume represent a collection of papers - about sedimentology.

I would like to thank those who took part in the IAMG session at the Congress. Prof. R.A. Reyment of Uppsala University and president of IAMG, kindly helped chair the session. I also would like to thank others who were unable to attend NICE but supplied papers for inclusion in these proceedings. Prof. Ph. J. Mangin of the University of Nice and Mr. Rene Bonnefille of

Electricite de France assisted with physical arrangements at the Congress. Mrs. Janice Potak and Mrs. Denise Jud of the Department of Geology, Syracuse University typed the manuscripts and ably proofed the gallies. Mr. Harry Holt of Pergamon Press arranged for publication.

9 July 1975
Nice, France

D.F. Merriam
Department of Geology
Syracuse University
Syracuse, New York (USA)

# AN ANALYSIS AND MANAGEMENT SYSTEM SUITABLE FOR SEDIMENTOLOGICAL DATA

John M. Cubitt[1]

*Syracuse University*

## ABSTRACT

G-EXEC is a data-analysis and management system developed since 1972 by the Institute of Geological Sciences (IGS). System concepts incorporated in its design, including integration, modularity and generalization, enable the system to be (1) transported from computer to computer, (2) updated rapidly, (3) operated on any data set, and, of primary importance, (4) used by noncomputer-oriented sedimentologists. Control of the system by the user proceeds through near-English commands which are translated into FORTRAN by the G-EXEC controller. The implementation of G-EXEC at the IGS enables the sedimentologist to employ programs for organization, manipulation, retrieval, statistical analysis and display of the data. KEY WORDS: *data systems, general, sedimentology*.

## INTRODUCTION

Sedimentological data files form a large proportion of the geological data stored and retrieved by computer (Hubaux, 1972; Robinson, 1975). Computer-based systems for the analysis and management of these data are, however not widely known or documented (Burk, 1973). It seems appropriate at this symposium, therefore, to partially alleviate this problem by presenting a brief account of one data management and analysis package, G-EXEC, that has been applied extensively to sedimentological data, and by stimulating interest in the use of the system in other geological fields. This paper complements a presentation made in the COGEODATA state-of-the-arts review on computer-based systems for geological field data (Jeffery and Gill, 1973).

---

[1]Formerly: Computer Unit, Institute of Geological Sciences, Exhibition Road, London, UK.

G-EXEC (Geologists EXECutive) was developed in 1972 by the Computer Unit at the Institute of Geological Sciences for the management and analysis of data bases (Jeffery and others, 1974). As an integrated generalized data-handling system, G-EXEC has been applied to a wide variety of data files ranging through geochemical data, mineral-resource data, paleontological data, petrological data and administrative records, but the major areas of application have been sedimentology, field geology (Jeffery and Gill, 1973) and borehole records. This paper provides the sedimentologists with a basic outline of the system and discussion on the implementation and application in the field of sedimentary geology.

## SYSTEM DESIGN AND CONCEPTS

The basic principal in the design of G-EXEC is to facilitate the use of computers by geologists and other scientists. Logically, computer applications should be an extension of the scientists thought processes and to this end, the "proccss programs" of G-EXEC perform simple logical tasks which are identifiable as steps in geological reasoning and can be accessed through simple near-English commands. The units can be assembled in any combination or order to perform management or analysis of geological data, that is the user is not restricted to one chain of processes but may choose his own chain from a large group of processes. The concept of an integrated system means that the system, under new command, should be able to perform any process at any position in the chain of processes chosen by the user. Such a design philosophy demands that a standard communication medium is passed from process to process. This medium is a standard data set termed the G-STAR standard file and described in detail later.

The concept of a generalized system was introduced in the design stage of G-EXEC as a prerequisite of any geological management and analysis package (Jeffery and Gill, in preparation). A generalized system has two main attributes, machine independence and data independence. Machine independence in G-EXEC is achieved by writing the software in ANSI FORTRAN IV, an internationally recognized high-level language and therefore supported on almost any computer. This enables G-EXEC to be transferred from one computer to another without a large overhead in program conversion. Data independence is achieved by having a medium to describe the data file to the system. Such a medium is termed the data description (Fig. 1). This enables the system to process any dataset of the standard form, whether the data describes rocks, boreholes, chemistry or other geological information, and to operate on any combination of files.

The data description consists of geological records (Fig. 2) containing information such as file name, field names, user code (for security purposes), field types, length of fields and upper and lower limits. Thus, the data with their description forms a standard file processable by any G-EXEC process program. This file, termed a G-STAR standard file, is independently self-describing, can be incorporated into a data bank, can provide an exchange

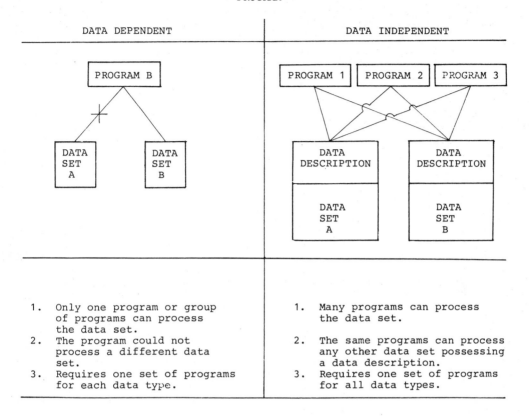

Figure 1. Data dependence versus data independence (after Jeffery and others, 1974).

format between programs and is suitable as an exchange format between scientists handling data.

## ORGANIZATION OF THE SYSTEM

The system is of modular design. The lowest level basic modules each carry out one clearly defined function (e.g. calculating correlation coefficients of a data set) and consist of a number of FORTRAN IV statements whereas macromodules are groups of modules performing a complete procedure (e.g. amending a record, a process involving reading from one device, comparing and writing to another device). Process programs are combinations of macromodules and modules which perform, for example, cluster analysis, data retrieval or simulation. Each of the process programs is placed for administrative convenience, into a package of programs that covers an area of operation such as display, data management or analysis. The system levels therefore are:

```
System
    Package
        Process Program
            Macromodule
                Module
                    FORTRAN Statement
```

DATA DESCRIPTION

| RECORD | CONTENT |
|--------|---------|
| 1. | File name/Number of fields in a record/Number of data records in a file/User code (for security) |
| 2. | Field names* |
| 3. | Field names* |
| 4. | Data record FORTRAN format |
| 5. | Ditto |
| 6. | Ditto |
| 7. | Type of field** |
| 8. | Length of each field** |
| 9. | Upper limit value for each field |
| 10. | Lower limit value for each field |
| 11. | Absent data code for each field |
| 12. | Names of dictionaries for each field |

  *Records 2 and 3 may occupy only one 80-column card but will be stored as two records.
  **Records 7 and 8 are read in together but are separated for storage.

Figure 2.  Content of data description.

The G-EXEC system is continually evolving, with replacement of modules, macromodules and process programs when improved algorithms become available and with addition of programs should new techniques develop.

The G-EXEC package (as distinguished from the G-EXEC system) provides the system controller and administrative software.  The controller arranges that the user commands are translated to appropriate FORTRAN statements (to invoke the required program

MANAGEMENT SYSTEM SUITABLE FOR SEDIMENTOLOGICAL DATA

modules) and Job Control Language to access the G-STAR standard files.

## APPEARANCE OF G-EXEC TO THE USER

The user controls the G-EXEC system by means of near-English commands that fall in a hierarchic order.

1.     SYSTEM COMMANDS
2.             EXECUTIVE COMMANDS
3.                         PROCESS COMMANDS.

1. System commands specify the user identification to the system and the overall requirements of time and output for a G-EXEC run. There are three such commands; (a) the systems parameter record, which provides the time and output requirements to the computer operating system; (b) the users identification record, which identifies the user to the G-EXEC system and provides security checks; and (c) the STOP command which terminates a G-EXEC run.

2. Executive commands cause the system to respond to the users requirements. There are only three such commands:

EXEC - instructs the system to prepare a process program for the user

FIND - instructs the system to prepare a file of data for the user

MAKE - instructs the system to write new data to a file.

3. Process program commands allow the EXEC command to control the functioning of each process program. The commands normally contain a combination of options records enableing the user to process and output data in a variety of manners.

A generalized example of the use of these commands is presented in Figure 3. The EXEC-options-FIND process can be repeated to complete multiple operations in a single job (Fig. 4).

Supplementing these simple processes, executive and system commands are five commands that are used with programs in the storage and retrieval package (G-STAR) of G-EXEC. The commands are WHEN, ALPH, and WORD which allow retrieval of numeric or alphanumeric data, HOLD which enables selected fields to be stored and END which completes a retrieval job. An example of the use of these commands is presented in Figure 5.

## IMPLEMENTATION

G-EXEC has been implemented at the Institute of Geological Sciences, London (IGS) to process and manage geological data. The

```
       Line
     number              Job control language cards
       1   G-EXEC SA001       10   5
       2   JOHN CUBITT, GEOLOGY DEPARTMENT, SYRACUSE UNIVERSITY        SA001    01-09-750001
       3   EXEC GCLUST         100    10
       4   1110000000           13  1
       5   FIND CARDFILE
                              DD follows here
                              Data follows here
       6   STOP
                         Job control language cards
```

Figure 3.  Example of cluster analysis by G-EXEC. Analysis of 100 samples by 10 variable data matrix (specified on line 3) using correlation coefficients (option = 1 in column 24 of line 2) involves entering systems parameter card (line 1) - showing project code (columns 8 to 12), time requirements (number of minutes in columns 19 and 20), and output limitations (thousands of lines output in column 23); user identification card (line 2) containing user name and address, project code (columns 61 to 65) for security purposes, date (columns 69 to 76) and personnel user code (columns 77 to 80); executive command card (line 3); options card (line 4); another executive command card (line 5) and finally job completion card (line 6). Surrounding G-EXEC command cards are job control language cards for accessing computer operating system and compilers.

system is installed on an IBM System/360 Model 195 at the Rutherford High Energy Laboratory (operated by the Science Research Council) and is accessed via a PDP 11/10, a remote job-entry station at the IGS.

Software totals greater than 30,000 lines of source code, split into packages for documentation convenience. The packages are G-EXEC, controller software; G-STAR, storage and retrieval facilities; G-UTIL, utility software; G-ANAL, analysis, and statistics programs; G-POST, postprocessing; G-PREP, preprocessing;

```
           G-EXEC
           JOHN CUBITT, SYRACUSE
           EXEC GCLUST              cluster analysis
             options
           FIND TESTDATA
           EXEC GFACT               factor analysis
             options
           FIND TESTDATA
           EXEC GRMLT2              multiple regression analysis
             options
           FIND TESTDATA
           STOP
```

Figure 4.  Generalized multiple operation G-EXEC job. Note that only basic details are included here and inevitably variety of extra data needs to be supplied for processing (see Fig. 3).

```
Line                Job control language cards
number
  1   G-EXEC SA001        10     5
  2   JOHN CUBITT,GEOLOGY DEPARTMENT,SYRACUSE UNIVERSITY       SA001    01-09-750001
  3   EXEC GRETSB              5000     20
  4    1
  5   (1X,5F10.4,3X)
  6   ( )
  7   ( )
  8   FIND TEST RETRIEVAL SET
  9   WHEN XCOORD    .GE. 10000.0 .LE. 25000.0
 10   WHEN YCOORD    .GE.  2500.0 .LE. 12500.0
 11   WORD COUNTRY   .EQ.SCOTLAND
 12   ALPH SAMPNUMB.GE.2453        .LE.9712
 13   HOLD COPPER   ,LEAD   ,SILVER  ,URANIUM ,ZINC
 14   END
 15   MAKE TEMPFILE
 16   STOP
                    Job control language cards
```

Figure 5.  Example of retrieval using G-EXEC. Values of copper, lead, silver, uranium and zinc present in geochemical samples 2453 to 9712 from area, bounded by X coordinates 10000.0 and 25000.0 and Y coordinates 2500.0 and 12500.0, in Scotland are retrieved and stored in temporary disc file. Lines 4, 5, 6 and 7 represent command for printing temporary file and formats of printed output.

G-PLOT, data display; G-CART, cartographic software; G-TRAN, transformation software; G-VET, vetting and validation software and G-SIM, simulation package.

The IGS implementation, at present has some 250 files ranging through field geology, borehole records, sedimentology, geochemical field data, geochemical analysis data, bulk mineral-resource data, geotechnical data, mineral production and trade statistics, paleontology, petrology, hydrogeology, structural geology, and administrative data. Files range to 10,000 records in length with a record length up to 500 bytes. This system is accessed by some 40 users and handles up to 700 G-EXEC jobs a month.

The system also is implemented at IGS Edinburgh on a PDP 11/45. Smaller versions of the system are available at the Norwegian Geological Survey, Trondheim; University of Uppsala, Sweden, Geological Survey of Sweden; German Geological Survey, Hannover; Spanish Geological Survey, Madrid; The Netherlands Soil Survey; the Institute of Hydrology, England; University of Costa Rica; and several UK, Canadian and USA universities. Installation and systems maintenance has been the responsibility of the respective institutions. Reports on a number of these implementations are available from the computer unit at the IGS (the address for information is presented after the conclusions).

## PROCESS PROGRAMS

Applications of G-EXEC to sedimentological data have concentrated on several sections of G-EXEC software. Although G-EXEC has been employed to organize, manipulate and retrieve sedimentological data (G-STAR, G-UTIL), statistical analysis (G-ANAL) and display (G-PLOT) are the packages employed in most applications.

G-ANAL consists of process programs for the calculation of univariate, bivariate and multivariate statistics, including linear and polynomial regression, cumulative frequency curves and histograms, smoothing of data, equal spacing of data and Fourier analysis. A program is available for calculating Folk statistics, McCammon statistics and Folk textural description from an input of weights for specific PHI values and percentages of sand, clay and silt.

Multivariate statistical facilities available in G-EXEC include factor analysis, principal components analysis, cluster analysis, multiple regression, multiple discriminant analysis, nonlinear mapping and correspondence analysis (all are generally adapted from previously published programs). To supplement the statistical and management programs, a number of display-oriented programs (G-PLOT) have become available in G-EXEC. These include trend-surface analysis, contouring, double Fourier analysis, frequency distribution analysis and display, Piper diagram display, gray-scale mapping, scatter-plot display and triangular diagrams. At present all display programs are restricted to line-printer output although research into forming a generalized graph plotter interface is currently being conducted.

## CONCLUSIONS

G-EXEC is a continuously developing, generalized, data management and analysis system that is currently implemented in many universities, geological surveys, and government institutions around the world. The main features of the system are machine and data independence, modular design and controller translation of near-English commands. These enable the user to submit simple instructions to the system without the necessity of learning a programming language, to apply any program to his data and to perform a number of processes in a single job. A wide variety of data management facilities enable the user to organize, manipulate, retrieve and store his data. Statistical analysis and display programs developed for the analysis of geological data supplement the systems capabilities. All of these features make the G-EXEC system particularly suited to the analysis of both numeric and nonnumeric data, for example lithological descriptions and mineralogical percentage data, a feature common in sedimentary analysis. Although not primarily designed for any one section of geology, the system has considerable experience in sedimentological data analysis.

## AVAILABILITY

For further information about the system, contact your nearest G-EXEC representative. The "home" of G-EXEC is,

Dr. K.G. Jeffery
Computer Unit
Institute of Geological Sciences
Exhibition Road, South Kensington
London SW7 2DE
England

and information can be obtained from:

Dr. J.M. Cubitt
Department of Geology
Syracuse University
Syracuse, New York  13210  USA

Dr. P. Sutterlin
Department of Geology
University of Western Ontario
London, Ontario, Canada

## ACKNOWLEDGMENTS

Constructive suggestions and criticisms of the manuscript were provided by Drs. D.F. Merriam, O.B. Nye, and S. Henley. The author also wishes to thank Mrs. Janice Potak and Miss Alice Salisbury for typing the manuscript and preparing the diagrams.

## REFERENCES

Burk, C.F., Jr., 1973, Computer-based storage and retrieval of geoscience information: bibliography, 1970-1972: Geol. Survey Canada Paper 73-14, 38 p.

Hubaux, A., ed., 1972, Geological data files: Codata Bull. 8, 30 p.

Jeffery, K.G., and Gill, E.M., 1973, G-EXEC: a generalized FORTRAN system for data handling, *in* Computer-based systems for geological field data: Geol. Survey Canada Paper 74-63, p. 59-61.

Jeffery, K.G., and Gill, E.M., in preparation, Design philosophy of G-EXEC: Computers & Geoscience.

Jeffery, K.G., Gill, E.M., Henley, S., and Cubitt, J.M., 1974, G-EXEC system, user's manual: Inst. Geol. Sci.

Robinson, S.C., 1974, The role of a data base in modern geology, *in* The impact of quantification on geology: Syracuse Univ. Geol. Contr. 2, p. 67-81.

# TREND ANALYSIS OF SEDIMENTARY THICKNESS DATA: THE PENNSYLVANIAN OF KANSAS, AN EXAMPLE

K.P. Thrivikramaji and D.F. Merriam

*Syracuse University*

ABSTRACT

Trend analysis of sedimentary thickness data in areas of flatlying strata can give insight into regional variation and local changes of structure and structural development. An analysis of of the Pennsylvanian sequence in Kansas shows that the regional structure was influenced by development of tectonic elements outside the State and that local structure was a continued development of features controlled by an earlier imposed tectonic pattern. A "break" in structural pattern (albeit minor) was found between the Desmoinesian and Missourian. This change also is evident in the character of sediments and perhaps is a more important "break" than heretofore recognized. KEY WORDS: *cluster analysis, trend-surface analysis, Kansas, structure, structural development, sedimentary thickness data.*

INTRODUCTION

Geologists have long been intrigued with the use of trend analysis especially for use in structural studies (e.g., Merriam and Harbaugh, 1963, 1964). The technique also is extended easily to sediment thickness data in the study of the structural development of an area (Merriam and Lippert, 1964). In this situation the following assumptions are made (1) the marker horizons were essentially flat and horizontal at time of formation, and (2) the thickness interval represents the composite structural development during that time. Thus the variation in thickness may be visualized as structure on the lower surface at the time the upper one was flat and horizontal; thick areas will indicate synclines and thin areas depict anticlines. If several intervals are studied in sequence, much can be learned about the structural history of an area through time (Merriam and Lippert, 1966).

For this study we were interested in the development of regional structure during Pennsylvanian time in Kansas.

## Location of Area

Kansas is located in the Midcontinent region of the United States, which is on the Stable Interior, a southern extension of the Canadian Shield. Sedimentary cover forms a thin veneer over the Precambrian crystalline basement with a maximum thickness of about 10,000 ft in Kansas.

## Stratigraphy

These strata range in age from Cambrian to Recent. Mississippian and Pennsylvanian rocks are exposed along eastern margin of the State, whereas younger beds outcrop to the west; rocks older than Mississippian do not outcrop.

Pennsylvanian rocks in Kansas are well known for their cyclic nature, and much has been written about them (e.g., Moore, 1949; Merriam, 1963). They are divided into five series as shown in Table 1. The Permian-Pennsylvanian contact is conformable whereas the Pennsylvanian-Mississippian contact is unconformable. The beds dip gently to the west and form approximately north-south outcrop bands. Drilling for groundwater, oil, and natural gas has added much information on the subsurface of these rocks mantled by younger deposits.

Table 1. Subdivisions of Pennsylvanian rocks in Kansas

|  |
| --- |
| Virgilian |
| Missourian |
| Desmoinesian |
| Atokan |
| Morrowan |

## Source of Data

The data were interpreted from electric and gamma-ray logs recorded in wells penetrating Pennsylvanian rocks in Kansas. Stratigraphic information on 469 wells was used in this study (Fig. 1).

## TECHNIQUE

Trend-surface analysis is an extension of regression analysis to include geographical coordinates. Surfaces are fit to spatial data using the principle of least squares. This ensures that the sum of squares of the deviations from the fitted surface is minimum.

$$(Y_{obs} - Y_{trend})^2 = minimum$$

where,

$Y_{obs}$ = observed data value

$Y_{trend}$ = predicted (computed) value

General form of polynomials for higher degrees is

$$Y = b_1 X_1 + b_2 X_2 + b_3 X_3 + \ldots \ldots + b_m X_m$$

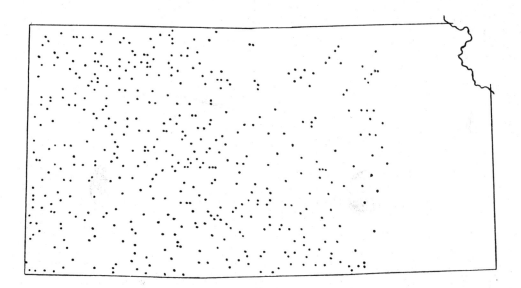

Figure 1 - Index map of Kansas showing data distribution.

Trend-surface analysis separates the data into two components, regional (trend) and local (residuals or deviations). Depending on the nature of the problem either the trend or residuals may be of interest. The residuals, the difference between the observed and computed values, are positive if the surface passes below the points and negative if above. It should be kept in mind that in studies involving thickness data the positives are thick and the negatives thin so that anticlines are represented by thin areas (negatives) and basins by thick areas (positives).

Statistical tests are used to determine how well the computed surface represents the actual data. Goodness of fit is an index of the total variation within the data expressed as a percentage. A perfect fit for a surface would yield 100 percent and anything less than a perfect fit will yield a proportionately less percentage value.

A second test used is the analysis of variance. If the regression is significant the variance associated with the deviations will be small in contrast to that associated with the trend function. The F-test tests if the variance could have resulted from random samples

## INTERPRETATION

To help interpret the structural history of Kansas during the Pennsylvanian, trend surfaces up to the third degree were fit to the subsurface thickness data. Because lower Pennsylvanian beds are restricted to the deepest parts of the basins only three intervals were recognized for this analysis. The interval labeled Desmoinesian includes Morrowan and Atokan units. Statistical data on the trend analysis are given in Table 2. All of the surface fits are significant at the 99.9 percent level.

Table 2. Statistical data on trend surfaces fitted to Pennsylvanian thickness data in Kansas

| Thickness interval | No. of points | Percent total sum of squares represented by each surface | | | Confidence level | | |
|---|---|---|---|---|---|---|---|
| | | 1st | 2nd | 3rd | 1st | 2nd | 3rd |
| Desmoinesian | 467 | 17.0 | 45.2 | 53.8 | 99.9 | 99.9 | 99.9 |
| Missourian | 467 | 32.7 | 42.7 | 45.6 | 99.9 | 99.9 | 99.9 |
| Virgilian | 467 | 85.9 | 89.8 | 91.0 | 99.9 | 99.9 | 99.9 |
| Desmoinesian-Missourian | 467 | 36.0 | 58.9 | 65.0 | 99.9 | 99.9 | 99.9 |
| Total Pennsylvanian | 467 | 57.7 | 79.6 | 84.0 | 99.9 | 99.9 | 99.9 |

### Structural Development

The regional slope of the area in early Pennsylvanian time was to the southwest (Fig. 2). There is a mild reflection of the positive Central Kansas Uplift complex in the central part of the State and associated lows. Pennsylvanian sediments were deposited on this eroded (in places karst), southwestern-sloping surface. As the basins filled, the sediments overlapped the positive structures until the surface was completely covered (Merriam, 1963). The masking effect of the sediments can be seen in the statistical data for the fits (Table 2). For example the fit of the first-degree surfaces increase from 17.0 percent (Desmoinesian) to 85.9 percent (Virgilian) reflecting less variation in the thickness values in the younger intervals.

During Missourian time the tilt of the area was increased to the south and east reflecting developing structures outside the Kansas area. The pre-Pennsylvanian structures were overlapped subduing the influence of these features although they continued to develop. This pattern continued through the Virgilian (Fig. 2). The change in slope from southwest (Desmoinesian) to south

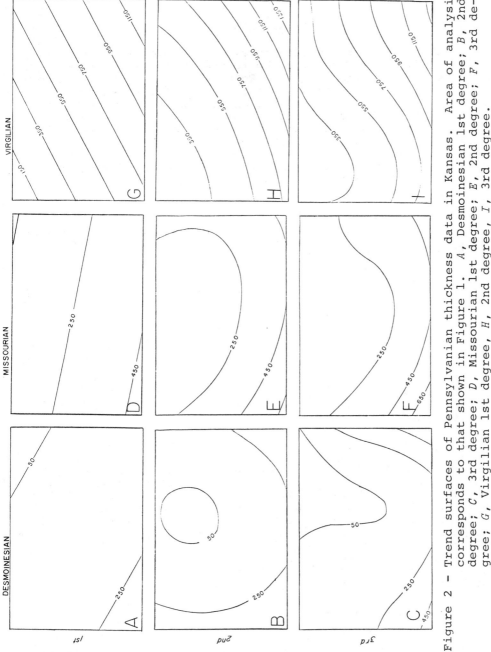

Figure 2 - Trend surfaces of Pennsylvanian thickness data in Kansas. Area of analysis corresponds to that shown in Figure 1. *A*, Desmoinesian 1st degree; *B*, 2nd degree; *C*, 3rd degree; *D*, Missourian 1st degree; *E*, 2nd degree; *F*, 3rd degree; *G*, Virgilian 1st degree, *H*, 2nd degree, *I*, 3rd degree.

(Missourian) and then southeast (Virgilian) can be attributed to downwarping of the Anadarko Basin to the south in Oklahoma and rise of the Ancestral Rocky Mountains to the west in Colorado.

The cumulative effect of the structural movement during the Pennsylvanian is shown in Figure 3. In general, the patterns show the change in configuration of the lower surface (top of Mississippian = base of Pennsylvanian) through time. The initial pattern imprinted during the Desmoinesian is emphasized during the remainder of the Pennsylvanian. The dominant regional feature is the Central Kansas Uplift complex. The residuals (not shown) indicate local structures, as defined during the Precambrian and lower Paleozoic, continued to develop also. The surfaces show an improvement in fit (not the first degree: Desmoinesian-Missourian, 36.0%; and total Pennsylvanian, 57.7%) reflecting the less variation in thickness as the initial roughness is smoothed by the blanket effect of the infilling sediments.

## Present Structure

Present structure is shown in Figure 4 and statistical data for the surface fits are summarized in Table 3. This configuration represents the total development in post-Pennsylvanian time and essentially is that of regional tilting.

The first-degree trends have a steeper slope on the older surfaces. This agrees with evidence from other sources that indicate structural complexity increases with depth or age of surfaces (Merriam, 1963). The slope changes (from oldest to youngest) from southwest to southeast reflecting the general eastward tilt of Kansas as a result of development of the basins to the south and the mountain ranges to the west.

The second- and third-degree surfaces show much the same picture — that of the positive Central Kansas Uplift complex (saddle) and adjacent basinal features (negative, low areas). A slight shift in the saddle southwards and eastwards is evident suggesting an overall tilt to the southeast through time. This interpretation is in accord with that made of the first degree surfaces.

Table 3. Statistical data on trend surfaces fitted to Pennsylvanian structural data in Kansas

| Present structure | No. of points | Percent total sum of squares represented by each surface | | | Confidence level | | |
|---|---|---|---|---|---|---|---|
| | | 1st | 2nd | 3rd | 1st | 2nd | 3rd |
| Mississippian | 469 | 45.5 | 74.4 | 76.3 | 99.9 | 99.9 | 99.9 |
| Desmoinesian | 469 | 23.4 | 69.3 | 74.3 | 99.9 | 99.9 | 99.9 |
| Missourian | 469 | 11.6 | 73.1 | 79.3 | 99.9 | 99.9 | 99.9 |
| Virgilian | 469 | 6.8 | 81.2 | 86.3 | 99.9 | 99.9 | 99.9 |

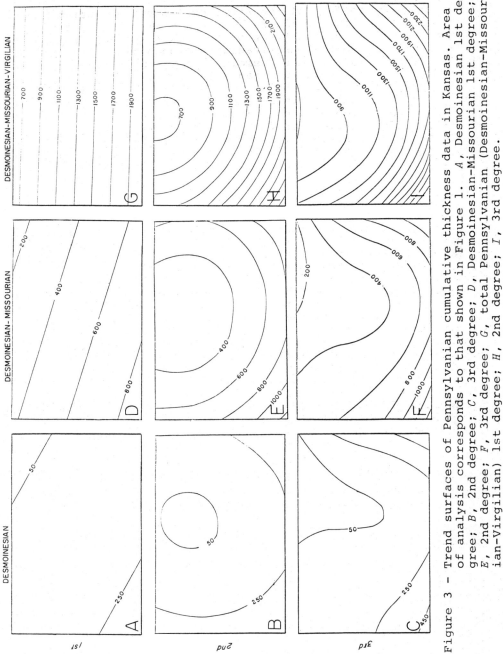

Figure 3 — Trend surfaces of Pennsylvanian cumulative thickness data in Kansas. Area of analysis corresponds to that shown in Figure 1. *A*, Desmoinesian 1st degree; *B*, 2nd degree; *C*, 3rd degree; *D*, Desmoinesian-Missourian 1st degree; *E*, 2nd degree; *F*, 3rd degree; *G*, total Pennsylvanian (Desmoinesian-Missourian-Virgilian) 1st degree; *H*, 2nd degree; *I*, 3rd degree.

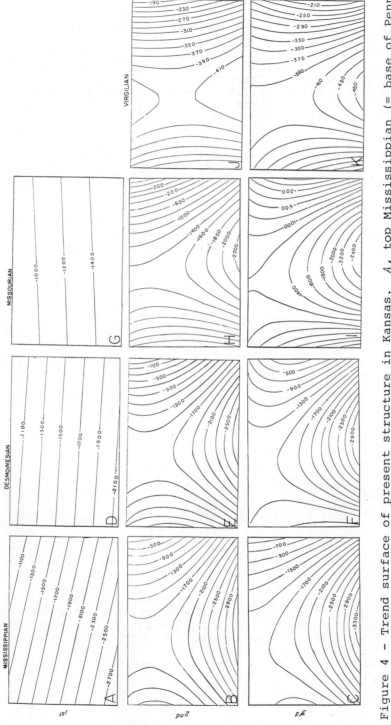

Figure 4 — Trend surface of present structure in Kansas. *A*, top Mississippian (= base of Pennsylvanian) 1st degree; *B*, 2nd degree; *C*, 3rd degree; *D*, top Desmoinesian 1st degree; *E*, 2nd degree; *F*, 3rd degree; *G*, top Missourian 1st degree; *H*, 2nd degree; *I*, 3rd degree; *J*, top Virgilian 2nd degree; *K*, 3rd degree. 1st degree fit on top of Virgilian not significant.

## Map Comparison

To determine the degree of similarity between the maps, a cluster analysis was performed on the trend coefficients. This is one of several techniques used to compare maps (Merriam and Sneath, 1966). Other techniques include differencing (Mirchink, and Bukhartsev, 1960), the computed trend matrices (Miller, 1964), trend residuals (Merriam and Lippert, 1966), shape parameters (Demirmen, 1972), vectorial descriptors (Rao and Srivastava, 1969), and spectral attributes (Rayner, 1967).

The trend coefficients are given in Table 4. The coefficients of the third-degree surfaces (9 terms) were used as numeric descriptors of the structural configuration. Correlation and distance coefficients were computed for input for the cluster analysis.

Table 4. Coefficients, excluding base values (A), of first-, second-, and third-degree trend surfaces fitted to structural data on top of Virgilian (V), Missourian (M), Desmoinesian (DM), and Mississippian (MI) of Kansas.

| | | B | C | D | E | F | G | H | I | J |
|---|---|---|---|---|---|---|---|---|---|---|
| V | 1st | 1.8039 | -0.6916 | | | | | | | |
| | 2nd | 14.2806 | -6.9058 | 0.0673 | -0.0188 | -0.0181 | | | | |
| | 3rd | 21.4972 | -8.1369 | 0.1055 | 0.0389 | -0.0351 | 0.0000 | 0.0003 | -0.0001 | -0.0000 |
| M | 1st | -0.2025 | -3.7270 | | | | | | | |
| | 2nd | 10.5752 | -11.9926 | 0.0645 | -0.0296 | -0.0217 | | | | |
| | 3rd | 18.7208 | -13.3414 | 0.1111 | 0.0240 | -0.0278 | 0.0000 | 0.0004 | -0.0001 | 0.0000 |
| DM | 1st | 0.2703 | -5.5495 | | | | | | | |
| | 2nd | 9.3870 | -17.9618 | 0.0549 | -0.0257 | -0.0439 | | | | |
| | 3rd | 15.8717 | -20.7845 | 0.0863 | 0.0163 | -0.0623 | -0.0000 | 0.0003 | -0.0002 | 0.0000 |
| MI | 1st | 1.6204 | -7.3096 | | | | | | | |
| | 2nd | 8.6942 | -19.3676 | 0.0404 | -0.0151 | -0.0484 | | | | |
| | 3rd | 14.5397 | -19.5051 | 0.0763 | 0.0391 | -0.0678 | 0.0000 | 0.0002 | 0.0000 | -0.0001 |

Results are shown in Figure 5. Both dendrograms are similar and both reinforce the visual interpretation. The younger surfaces (Virgilian and Missourian) are similar and the two older ones (Desmoinesian and Mississippian) are similar. The "break" (albeit minor) occurs between the Desmoinesian and Missourian. This structural "break" is reflected in a change in the sediments from a dominantly clastic sequence with coal beds to one of shales and interbedded carbonates largely of marine origin. The cyclothems also change in character as would be expected.

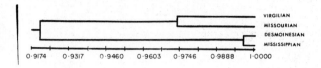

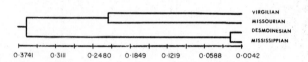

Figure 5 - Dendrogram of cluster analysis of 3rd-degree fit of structural data. Use of correlation coefficient r (up diagram), and distance coefficient, $d^2$ (lower diagram) gives similar results.

## SUMMARY

Trend-surface analysis can be used to study quantitatively the structure and structural development of areas of essentially flat-lying sedimentary sequences. Coupled with cluster analysis, the technique is effective in determining "breaks" in the sequence and emphasizing features not readily recognized by visual inspection of conventional graphics. This type of analysis also may point out areas for, or additional directions of, investigation.

Thickness of Pennsylvanian units in Kansas shows a progressive and continual structural development during Pennsylvanian time. A minor change in the pattern of development occurred between the Desmoinesian and Missourian. The Central Kansas Uplift complex and associated lows exerted considerable influence in the Pennsylvanian. The entire area was tilted southward as a result of the development of the Anadarko Basin in Oklahoma and eastward as a result of uplift of the Ancestral Rocky Mountains. Post-Pennsylvanian history was mainly one of tilting with structures developing locally as a result of continued adjustment of sediments over basement-controlled features.

## REFERENCES

Demirmen, F., 1972, Mathematical procedures and FORTRAN IV program for description of three-dimensional surface configurations: unpubl. rept., KOX Project, Kansas Geol. Survey, 131 p.

Merriam, D.F., 1963, The geologic history of Kansas: Kansas Geol. Survey Bull. 162, 316 p.

Merriam, D.F., and Harbaugh, J.W., 1963, Computer helps map oil structures: The Oil and Gas Jour., v. 61, no. 47, p. 158-159, 161-163.

Merriam, D.F., and Harbaugh, J.W., 1964, Trend surface analysis of regional and residual components of geologic structure in Kansas: Kansas Geol. Survey Sp. Dist. Publ. 11, 27 p.

Merriam, D.F., and Lippert, R.H., 1964, Pattern recognition studies of geologic structure using trend-surface analysis: Colorado Sch. Mines Quart., v. 59, no. 4, p. 237-245.

Merriam, D.F., and Lippert, R.H., 1966, Geologic model studies using trend-surface analysis: Jour. Geology, v. 74, no. 3, p. 344-357.

Merriam, D.F., and Sneath, P.H.A., 1966, Quantitative comparison of contour maps: Jour. Geophysical Res., v. 71, no. 4, p. 1105-1115.

Miller, R.L., 1964, Comparison-analysis of trend maps, *in* Computers in the Mineral Industries (pt. 2): Stanford Univ. Publ., Geol. Sci., v. 9, no. 2, p. 669-685.

Mirchink, M.F., and Bukhartsev, V.P., 1960, The possibility of a statistical study of structural correlations: Doklady Akad. Nauk SSSR (English trans.), v. 126, p. 495-497.

Moore, R.C., 1949, Divisions of the Pennsylvanian System in Kansas: Kansas Geol. Survey Bull. 83, 203 p.

Rao, S.V.L.N., and Srivastava, G.S., 1969, Comparison of regression surfaces in geologic studies: Kansas Acad. Sci. Trans., v. 72, no. 1, p. 91-97.

Rayner, J.N., 1967, Correlation between surfaces by spectral methods, *in* Colloquium on trend analysis: Kansas Geol. Survey Computer Contr. 12, p. 31-37.

# NUMERICAL CLASSIFICATION OF MULTI-VARIATE PETROGRAPHIC PRESENCE-ABSENCE DATA BY ASSOCIATION ANALYSIS IN THE STUDY OF THE MIOCENE ZIQLAG REEF COMPLEX OF ISRAEL

B. Buchbinder and D. Gill

*Geological Survey of Israel*

ABSTRACT

Objective facies discrimination of the relatively homogeneous biocalcarenites of the Ziqlag Formation was accomplished by means of two numerical classification procedures suitable for handling multivariate qualitative data, cluster analysis and association analysis. The latter resulted in more homogeneous groups and, on the basis of comparisons with Recent assemblages of the same genera, provided a more meaningful classification of paleoenvironments. Coral reef, back-reef platform, and beach facies were recognized, each with a number of auxiliary subfacies. KEY WORDS: *association analysis, classification, carbonate petrology, paleoenvironments.*

INTRODUCTION

Remnants of the Upper Miocene Ziqlag Formation occur as isolated outcrops along the western foothills of the Judea-Samaria Mountains and as scattered patches in the subsurface of the Mediterranean coastal plain of Israel (Gvirtzman and Buchbinder, 1969; Gvirtzman, 1970; Buchbinder, in preparation). The formation consists essentially of biocalcarenites and, except for a few locations where a genuine coral-biolithite reef-core is exposed (Buchbinder, 1969), a detailed differentiation of facies is exceedingly difficult. The problem is aggravated wherever subsurface samples are concerned and it was, therefore, decided to examine the usefulness of multivariate methods to resolve some of the difficulties outlined.

The acquisition of accurate quantitative data for this type of analysis entails point counting or some other rigorous form of modal analysis. In addition to the tedium involved, the exclusive adherence to precise quantitative data imposes, as noted by

Bonham-Carter (1965, p. 483-484), several other undesired limitations on the analysis. Bonham-Carter (1965, 1967a, 1967b) demonstrated that satisfactory results also can be obtained from qualitative and semiquantitative data. Erez (1972) advanced one step further in showing that the power of discrimination realized by utilizing only the readily obtainable presence-absence data is, for all practical purposes, entirely adequate. The methodological ramifications of these successful case studies, namely a significant simplification in data acquisition and the ability to incorporate nonparametric data in the analysis, are attractive. However, particularly with reference to the association analysis method employed by Erez, "one swallow does not a summer make", and the examination of additional test cases seems in order before the method can be recommended for general practice in geology. Performing such a test constitutes an additional objective of this study.

## PROCEDURE

Thin sections of 249 surface samples were selected for the analysis. Using Terry and Chilingar's (1955) reference charts, visual estimates of the volume percent of each of the following 21 biogenic and rock constituents were obtained (the abbreviations, in parentheses, denote the mnemonic codes for the constituents as they appear in Fig. 1): Mollusks (Molk); Echinoids (Echd); Corals (Corl); Bryozoans (Bryo); *Halimeda* (Halm); Crustose coralline algae (Alga); *Amphiroa* (Amph); *Corallina* (Clna); Peneroplids including *Peneroplis*, *Spirolina* and *Dendritina* (Penp); *Borelis* (Bors); *Archai* (Arcs); Agglutinated foraminifers (Aggs); Miliolids (Mill); *Ammonia* Amon); *Elphidium* (Elph), *Amphistegina* (Amtg); *Operculina* (Oper); Other rotaliform foraminifers, including *Pararotalia*, *Asterigerina*, *Nonion*, *Cancris* and others (Rotl); Encrusting foraminifers (Encr); Quartz grains (Quaz) and Lithoclasts (Lcls).

The frequency distribution of each constituent was expressed in terms of the three modal classes of absent, present, and abundant, the mean volume-percent being the boundary between the last two classes. The three modal classes were coded into two two-state characters according to the additive scheme (Sokal and Sneath, 1963, p. 76). Excess weight was ascribed to the benthonic foraminifers and to *Amphiroa* by employing a third two-state character in coding their modal classes (Proctor and Kendrick, 1963). These semiquantitative data were subjected to cluster analysis, using program CLUST 3 (Bonham-Carter, 1967b). The same data set, reduced to the form of presence/absence data, also was subjected to association analysis, using program ASSOCA (Gill, Boehm, and Erez, in preparation). Only the results obtained by the latter are presented here.

## ASSOCIATION ANALYSIS

Association analysis is a numerical classification method developed by plant ecologists to evaluate qualitative, presence/absence type, multivariate data (Williams and Lambert, 1959, 1960,

1961; Lance and Williams, 1965; Noy-Meir, Tadmor and Orshan, 1970). Little use of this method has been made by geologists. Reference to it was made by McCammon (1968, p. 2181) but the only example of its application to geological problems known to us is in the study of the Recent depositional environments of the Gulf of Eilat (Erez, 1972).

In scope and purpose, the method is similar to the more familiar cluster analysis. Both produce a dendrogram portraying a hierarchical classification of either samples or variables. However, in their underlying strategies, the methods are different. In cluster analysis, the hierarchical tree is constructed from the "twigs downward" by progressively pooling individuals into groups based on overall similarities with respect to all measured attributes (i.e. "agglomerative-polythetic" strategy). In association analysis, the dendrogram is built from the "trunk upward" by dicotomously dividing already established groups, each division being based on a single attribute (i.e. "divisive-monothetic" strategy).

Similar to all other multivariate techniques, the algorithm evolves from a matrix of all pair-wise intervariable (R-mode, in terms of which the algorithm will be described) or intersample (Q-mode) similarities. The measure of association between any two variables (i) and (j) is the chi-square statistic, computed over the entire population by

$$\chi^2_{ij} = ((NN_{ij} - N_i N_j)^2 N)/(N_i(N-N_i)N_j(N-N_j))$$

where $N$ = total number of samples; $N_i$, $N_j$ = number of samples in which variables (i) and (j) are present, respectively, and $N_{ij}$ = number of samples in which both variables are present. With Yates' continuity correction, the coefficient is computed (Dixon and Massey, 1957, p. 226) by

$$\chi^2_{ij} = ((|NN_{ij} - N_i N_j| - N/2)^2 N)/(N_i(N-N_i)N_j(N-N_j))$$

The association parameter can range from 0 to $N$ = the number of samples in the entire population or in any subsequent group. These extreme values are obtained under the following conditions: if either or both variables are present or absent in all samples, the parameter is indeterminate and assumes the value of zero. Obtaining such a situation is, in effect, the objective of classification, because a group in which all intervariate associations are indeterminate (or insignificant) is homogeneous (Goodall, 1953). Chi-square assumes its maximal value if the variables (i) and (j) either mutually occur together, or are exclusive of each other. A division based on such an associative variable into two groups such that it is possessed by all samples of one and lacked by all samples of the other will render the associativity of this variable indeterminate in each of the resulting groups and, therefore, will yield two more homogeneous groups.

The row sums $\sum_{j=1, \neq i}^{m} \chi^2_{ij}$, summing the association coefficients of variables (i) with all other (m-1) variables, provide a measure of the overall associativity of the variables in the system. The most associative variable then is selected to divide the population into groups as described. The process is repeated in a similar manner for each of the new groups. At each successive division, the newly defined subgroups become progressively more homogeneous. The procedure is terminated when a preassigned level of homogeneity is reached. Such a precaution, which, incidentally, can not be employed in the agglomerative methods, guards against the creation of spurious groups arising from sampling errors or abnormal situations. Several measures of homogeneity and related termination rules have been proposed. Seemingly, none is entirely adequate (Lance and Williams, 1965; Lambert and Williams, 1966; Noy-Meir, Tadmor, and Orshan, 1970). The three more commonly recommended measures of homogeneity, the maximal single entry in the mxm matrix of chi-square values ($(\chi^2_{ij})$max; $i \neq j$) the sum of all the different entries in the matrix ($\frac{1}{2} \sum_{i=1}^{m} \sum_{j=1, \neq i}^{m} \chi^2_{ij}$) and the maximal row sum (($\sum_{i=1, \neq i}^{m} \chi^2_{ij})_{max}$) have been incorporated into our program as user's options. If the chosen parameter drops below a preassigned (optional) threshold value subdivision is terminated. Although the probabilistic validity of the test is subject to reservations, the threshold value most commonly used is 3.84, corresponding to $\chi^2$ (1, 0.05). Additional user's options pertain to some of the more specific aspects of the method which will not be discussed here. These include the application of Yates' correction to the association coefficient, discarding insignificant associations (i.e. $\chi^2_{ij} < 3.84$) from all computations and the treatment of ambiguities — the choice from among variables of equal associativity.

The resulting hierarchical classification is portrayed by a dendrogram whose dimensioned axis is scaled by a parameter chosen to express homogeneity (Fig. 1). If the final grouping obtained is too detailed for practical appraisal, the classification can be interpreted at any higher level of heterogeneity deemed appropriate for the situation at hand. Each final group is characterized by a positive (always present) and a complementary negative (always absent) suite of variables. This property can be used to advantage in classifying samples of unknown affinity from the same population which did not participate in the analysis. In addition, the dividing variables, occupying the dendrogram's nodes, facilitate an appraisal of the relative discriminative importance of the variables to the classification. However, because one or more variables might serve as dividers several times, at different levels and branches of the hierarchy, such an immediate ranking is usually valid only through the first three or four nodes. A comprehensive ranking for all the variables can be obtained from the frequencies of occurrence of the variables in the final groups, as suggested by Bonham-Carter (1967a, p. 577). Through the employment of a cutoff value to encode measurements into binary data, association analysis can

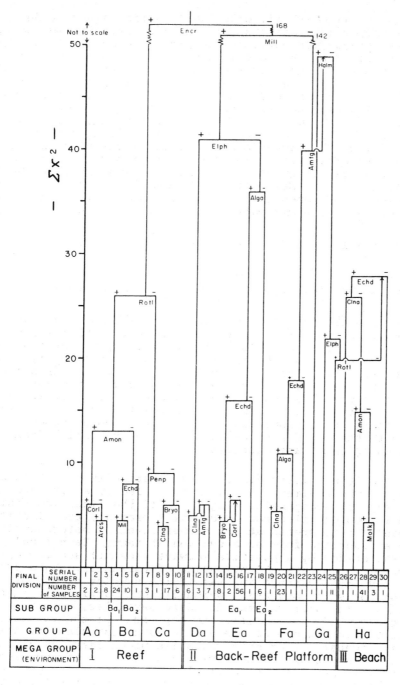

Figure 1. Dendrogram portraying hierarchical classification produced by association analysis of presence/absence data of 21 rock constituents in 249 samples from Ziqlag Formation.

be applied to quantitative data as well.  By transposing the original data matrix, the method can operate in the Q-mode to yield a grouping of variables rather than samples.

RESULTS AND DISCUSSION

The sample classification produced by the association analysis is summarized in the dendrogram shown in Figure 1. The options employed in this run included the application of Yates' correction: discarding insignificant associations; use of $\Sigma\chi^2$ as termination parameter and 3.84 as termination condition. The final division into 30 groups is clearly too meticulous to enable a practical interpretation. However, at a deliberate and controllable loss of internal homogeneity, geologically meaningful groups can be selected. Thus, for practical purposes, the final grouping was narrowed to eight groups (of differing degrees of internal homogeneity) which, at yet a higher level in the hierarchy, merge into three divisions or "megagroups" (Fig. 1). On the basis of their constituent composition and from supplementary field and petrographic data, the megagroups defined three mappable facies, representing reef, back-reef platform, and beach environments. The more detailed groups represent subfacies within these major environments. Because mollusks, echinoids, and coralline algae are ubiquitous, foraminifers are the most potent discriminators in the system. Furthermore, as all of the Upper Miocene foraminifers present are living today, the environmental significance of the various assemblages delineated by the analysis can be appraised by direct comparisons with Recent assemblages.

To better appreciate the distinctiveness of the resultant groups in terms of their constituent composition, one has to consult the list of samples arranged according to their dendrogram order, which because of space limitations could not be included here. The presence of encrusting foraminifers effectively separates the reef samples from all the rest. These foraminifers, more than corals or bryozoans, proved to be the best single diagnostic element in this megagroup.

Group Aa is characterized by a miliolids-rotaliforms-*Ammonia* assemblage with a frequent occurrence of *Amphistegina* and *Borelis*. The group does not represent a typical reefal buildup but rather the development of minor coral colonies in isolated patches within the back-reef platform. Group Ba is characterized by an agglutinants-rotaliforms assemblage, frequently with *Amphistegina* and group Ca by an agglutinants-*Amphistegina* assemblage. In addition to miliolids which are ubiquitous to groups Da and Ea, and rotaliforms, which are abundant in both groups, *Elphidium* is diagnostic of group Da whereas agglutinants and *Amphistegina* characterize group Ea. Group Ea, the largest group present, contains the largest assortment of foraminifers. This diversity in genera indicates that, within the back-reef platform, this group represents the most nonrestricted environment present. Groups Fa and Ga, characterized respectively by *Amphistegina* and rotaliforms, mark a gradual disappearance of the foraminiferal constituents and thus a progressive proximity to the shore (Buchbinder, in preparation). Group Ha is almost totally devoid of foraminifers. It consists almost exclusively of abraded fragments of mollusks, echinoids and coralline algae with a relative high frequency of quartz grains and lithoclasts. Its samples are located along the easternmost, landward, extensions of the formation and it most probably represents a beach facies. Whereas the reef and back-reef platform facies, for the most part, could be discerned by other means, the recognition of the beach facies and all of the more detailed subfacies is a direct contribution of the multivariate analysis.

The relative importance of individual constituents to the classification was computed from contingency tables constructed according to the selected groups. In decreasing order, the first ten were found to be encrusting foraminifers, miliolids, rotaliforms, *Elphidium*, *Amphiroa*, *Amphistegina*, *Ammonia*, agglutinants, bryozoans and corals. Only few of these have to be examined in order to allocate unknown samples into their respective groups. From the dendrogram, it is apparent that the first six groups (Aa to Fa) are uniquely defined by different combinations of three out of the first seven constituents. The ambiguity between the remaining two groups also will be resolved rather early if the sample contains any foraminifers at all.

The cluster analysis performed on the semiquantitative data resulted in a similar classification. In terms of identically classified samples, the agreement between the two methods is 89 percent at the major facies level and 56 percent at the more detailed group levels. Cluster analysis, being polythetic, is more tolerant to the inclusion of variables which are of no direct relevance to the classification and which, on the other hand, might lead a monothetic method astray. In the Ziqlag Formation, diagenetically affected components such as sparite and micrite, which did not significantly effect the results obtained by cluster analysis, had to be excluded from the association analysis in order to produce meaningful results. However, after this preparatory screening, the results obtained from the association analysis were more adequate in that its groups were more homogeneous and thus, more diagnostic in terms of paleoenvironments.

## ACKNOWLEDGMENTS

This study constitutes part of a doctoral dissertation by B. Buchbinder at the Hebrew University, Jerusalem, under the supervision of Drs. Z. Reiss and R. Rezak. Sincere thanks are extended to Dr. E. Noy-Meir and Mr. Y. Erez, whose computer program for association analysis served us during the early stages of this study; to S. Boehm, for programming and technical assistance; to Mrs. D. Ashkenazi, for typing the manuscript; and to the Geological Survey of Israel, under whose auspices this study was carried out.

## REFERENCES

Bonham-Carter, G.F., 1965, A numerical method of classification using qualitative and semi-quantitative data as applied to the facies analysis of limestone: Bull. Canadian Petrol. Geology, v. 13, no. 4, p. 482-502.

Bonham-Carter, G.F., 1967a, An example of the analysis of semi-quantitative petrographic data: Proc. 7th World Petroleum Cong., v. 2, p. 567-583.

Bonham-Carter, G.F., 1967b, FORTRAN IV program for Q-mode cluster analysis of nonquantitative data using IBM 7090/7094 computers: Kansas Geol. Survey Computer Contr. 17, 28 p.

Buchbinder, B., 1969, The Ofaqim reef complex - Ziqlag Formation: Geol. Survey Israel, Rept. OD/4/69, 16 p.

Buchbinder, B., in preparation, Lithogenesis of Miocene reef limestone in Israel with particular reference to the significance of red algae: unpubl. doctoral dissertation, Hebrew University, Jerusalem.

Dixon, W.J., and Massey, F.J., Jr., 1957, Introduction to statistical analysis: McGraw-Hill Book Co., Inc., New York, 488 p.

Erez, Y., 1972, Multivariate analysis of biogenic constituents in the sediments off Ras Burka, Gulf of Eilat: unpubl. masters thesis, Hebrew University, Jerusalem (in Hebrew), 58 p.

Gill, D., Boehm, S., and Erez, Y., in preparation, A FORTRAN IV program for R- and Q-mode association analysis with printed dendrograms.

Goodall, D.W., 1953, Objective methods for the classification of vegetation. I. The use of positive interspecific correlation: Australian Jour. Botany, v. 1, no. 1, p. 39-63.

Gvirtzman, G., 1970, The Saqiye Group (Late Eocene to Early Pleistocene) in the Coastal Plain and Hashefela regions, Israel: Geol. Survey Israel, Rept. OD/5/67 (in Hebrew), 170 p.

Gvirtzman, G., and Buchbinder, B., 1969, Outcrops of Neogene formations in the central and southern Coastal Plain, Hashefela and Be'er Sheva regions, Israel: Geol. Survey Israel Bull. 50, 73 p.

Lambert, J.M., and Williams, W.T., 1966, Multivariate methods in plant ecology. VI. Comparison of information analysis and association analysis: Jour. Ecology, v. 54, no. 3, p. 635-664.

Lance, G.N., and Williams, W.T., 1965, Computer programs for monothetic classification ("association analysis"): Computer Jour., v. 8, no. 3, p. 246-249.

McCammon, R.B., 1968, Multiple component analysis and its application in classification of environments: Am. Assoc. Petroleum Geologists Bull., v. 52, no. 11, pt. 1, p. 2178-2196.

Noy-Meir, I., Tadmor, N.H., and Orshan, G., 1970, Multivariate analysis of desert vegetation. I. Association analysis at various quadrant sizes: Israel Jour. Botany, v. 19, no. 4, p. 561-591.

Proctor, J.R., and Kendrick, W.B., 1963, Unequal weighting in numerical taxonomy: Nature, v. 197, no. 4868, p. 716-717.

Sokal, R.R., and Sneath, P.H.A., 1963, Principles of numerical taxonomy: W.H. Freeman and Co., San Francisco, 359 p.

Terry, R.D., and Chilingar, G.V., 1955, Summary of "Concerning some additional aids in studying sedimentary formations" by M.S. Shvetsov: Jour. Sed. Pet., v. 25, no. 3, p. 229-234.

Williams, W.T., and Lambert, J.M., 1959, Multivariate methods in plant ecology. I. Association analysis in plant communities: Jour. Ecology, v. 47, no. 1, p. 83-101.

Williams, W.T., and Lambert, J.M., 1960, Multivariate methods in plant ecology. II. The use of an electronic digital computer for association analysis: Jour. Ecology, v. 48, no. 3, p. 689-710.

Williams, W.T., and Lambert, J.M., 1961, Multivariate methods in plant ecology. III. Inverse association analysis: Jour. Ecology, v. 49, no. 3, p. 717-729.

# AN ASSESSMENT OF SOME QUANTITATIVE METHODS OF COMPARING LITHOLOGICAL SUCCESSION DATA

William A. Read

*Institute of Geological Sciences*

## ABSTRACT

Particular depositional environments are characterized by particular lithologic sequences. The overall comparison of lithologic successions is therefore a critical aspect of basin analysis. Few quantitative techniques have been developed however for making such overall comparisons.

Crossassociation similarity coefficients are only partially effective in revealing the sedimentologic relationships between fifteen borehole sections through Namurian deltaic deposits in the Kincardine Basin in central Scotland. The comparison of vectors of transition probabilities provides results which can be interpreted more easily, mainly because they can be related to a conceptual sedimentologic model. Product-moment correlation coefficients and Euclidean distances were calculated between each pair of vectors, and the results were classified by average-linkage cluster analysis. A nonlinear mapping program, which also uses Euclidean distances, presented the results in a more effective and readily interpretable manner. KEY WORDS: *crossassociation, nonlinear mapping, transition probabilities, sedimentology.*

## INTRODUCTION

Since the late 1940's, sedimentologists have become increasingly interested in regional analysis of sedimentary successions and particularly in basin analysis (e.g. Krumbein, 1948; Goodlet,

---

[1]Published by permission of Director, Institute of Geological Sciences.

1957; Allen, 1959; Potter and Pettijohn, 1963; Kelling, 1968; Friend and Moody-Stuart, 1972). Recently, the subject has been stimulated further by discovery that this line of research can assist in locating hydrocarbons (Fisher and McGowen, 1967, 1969; Fisher and others, 1969; Guevara and Garcia, 1972).

Following the lead of Allen and Krumbein (1962) and Duff and Walton (1964), the author and his colleagues have applied a series of quantitative techniques to the analysis of the Kincardine Basin in central Scotland, which was subsiding throughout Late Carboniferous time. Regional variations in bulk lithology were studied by a combination of lithofacies analysis and trend-surface analysis (Read, 1961, 1965; Read and Dean, 1967; Read, Dean and Cole, 1971), and the relationships between lithostratigraphic variables were studied by factor analysis and principal component analysis (Read and Dean, 1968, 1972).

A much clearer insight into the processes of sedimentation and depositional environments however can be obtained if the sedimentary successions are expressed in terms of strings of discrete lithologic states rather than overall bulk lithologies (Visher, 1965). This approach also enables the conceptual sedimentologic models, which have been derived from studies of modern depositional environments, to be linked to mathematical models (e.g. Read, 1969b; Johnson and Cook, 1973). It is advantageous to employ mathematical techniques to make quantitative overall comparisons between lithologic successions. Such techniques can be used, for example, to group similar sections and thus help to identify the depositional processes. Few workers however have investigated the quantitative comparison of strings of discrete lithologic states (as opposed to continuous variables) and more work will need to be done in this field before mathematical basin analysis can be fully developed.

Crossassociation is the technique that has been most used to date. It is analagous in some respects to time-trend analysis (Vistelius, 1961; Fox and Brown, 1965; Dean and Anderson, 1974), but compares strings of nonnumeric states instead of continuous variables. Crossassociation was developed originally to compare amino-acid chains and has had considerable success in this role in identifying matching sequences, breaks, and reversals (Sackin and Sneath, 1965). The technique also can be used to make overall comparisons between strings of nonnumeric data by calculating crossassociation similarity coefficients. In geology it has been used to correlate lithologic successions and to make overall comparisons between them (Sackin, Sneath, and Merriam, 1965; Merriam and Sneath, 1967; Read and Sackin, 1971); only its performance in the latter role is assessed in this paper.

An alternative approach was developed by Read and Merriam (1972). Transition-probability matrices provide an excellent summary of salient features of any lithologic succession which is formed by a repetition of the same basic sedimentary process (de Raaf, Reading and Walker, 1965; Selley, 1970) and also provide a basis of applying Markov-process models to the succession (Vistelius, 1949; Allègre, 1964; Carr and others, 1966; Potter and Blakely,

1967; 1968; Krumbein, 1967; Schwarzacher, 1967, 1969; Read, 1969b). In this new approach a vector of first-order transition probabilities is calculated for each lithologic succession, similarity coefficients are calculated between each pair of vectors, and a classification procedure is applied to the resulting similarity coefficients.

The aim of this presentation is to compare the results obtained by applying crossassociation and variants of the Read and Merriam (1972) technique to strings of lithologies which have been taken from a succession and a depositional basin where the sedimentology has been systematically investigated. Thus the results can be assessed against the background of existing sedimentologic knowledge. Hopefully the techniques which provide the most readily interpretable results for this specimen data set will be the most likely to provide satisfactory results when applied to successions and basins where the depositional environments and the tectonic frameworks are less well understood.

In order to keep this paper reasonably short, material which has been published in earlier papers by Read and Sackin (1971) and Read and Merriam (1972) is not repeated, and the reader is referred to those and to the paper on nonlinear mapping by Henley (1972) for additional details.

## DATA SET

The basic data in the public files of the Edinburgh Office of the IGS are the same as those used in the Read and Sackin (1971, list 2) crossassociation study and the Kincardine Basin data set used in the Read and Merriam (1972) study. A complete listing has been given elsewhere by the author (Read, 1970, table 34). The data were derived from geologists' decriptions of exactly the same stratigraphic interval in 15 cored boreholes drilled in the Kincardine Basin in central Scotland. The boreholes and numbers used to designate them are listed in Table 1 and located in Figure 1. The stratigraphic interval is the upper part of the Limestone Coal Group (Namurian, Pendleian Stage), between the widespread marine transgressions marked by the Black Metals and the Index Limestone (Read, 1959). The sequence is coal-measures facies and is known to have been deposited in a deltaic environment (Read, 1969a, p. 332; 1969b, p. 214-215).

The lithologic data were coded into five states, mudstone, siltstone, sandstone, seatrock, and coal. This level of coding into five states emphasized the underlying nonrandom nature of the data most clearly (Read and Sackin, 1971, p. 12). As in the earlier study by Read and Merriam (1972), no account was taken of the thickness of each bed, no lithology was allowed to pass upwards into itself, and only first-order transition probabilities were considered.

Table 1. List of boreholes from which basic data were obtained, together with numbers used to designate them and British National Grid References of sites (in brackets).

| | |
|---|---|
| A1 | Powis Mains No. 1 Bore, 1959 (NS 822958) |
| A2 | Torwood No. 1 Bore, 1927-28 (NS 835843) |
| A3 | Torwood Bore, 1960-61 (NS 838849) |
| A4 | Glenbervie No. 4 Bore, 1926 (NS 850857) |
| A5 | Tullibody No. 2 Bore, 1934-36 (NS 869938) |
| A6 | Doll Mill Bore, 1955 (NS 875881) |
| A7 | Mossneuk Bore, 1950-52 (NS 872861) |
| A8 | South Letham No. 1 Bore, 1952 (NS 886853) |
| A9 | Kincardine Bridge Bore, 1952-53 (NS 917872) |
| A10 | Orchardhead Bore, 1956 (NS 924841) |
| A11 | Grangemouth Dock Bore, 1956-57 (NS 951839) |
| A12 | Righead Bore, 1953-56 (NS 972882) |
| A13 | Culross No. 2 Bore, 1957 (NS 983859) |
| A14 | Solsgirth Bore, 1941-61 (NS 997948) |
| A15 | Shepherdlands Bore, 1933-34 (NT 006902) |

## METHOD

The process of calculating the crossassociation similarity coefficient values and of applying cluster analysis to these values was described in some detail by Read and Sackin (1971, p. 6-7). In the earlier study by Read and Merriam (1972), product-movement correlation coefficients were calculated between the vectors of transition probabilities for each pair of sections. Some doubt has been cast on the validity of correlation coefficients as similarity measures in a taxonomic role (see Everitt, 1974, p. 53-54), but Sneath and Sokal (1973, p. 140) consider that they are usually suitable. They therefore have been retained in part of this study.

A large number of alternative similarity coefficients could have been used (see Sneath and Sokal, 1973, p. 116-147), although it was impossible in the time available to investigate all of them. It was desirable nevertheless to check the robustness of the classification obtained by Read and Merriam (1972) by at least one parallel study using a different similarity coefficient,

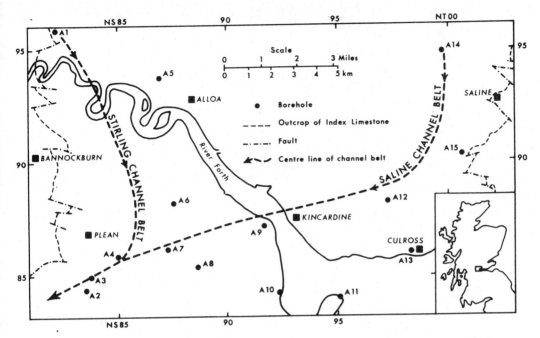

Figure 1. Sketch map of area studied showing sites of boreholes and center lines of channel belts.

and Euclidean distance was selected in view of its fundamental simplicity and widespread usefulness (Sneath and Sokal, 1973, p. 147). When employed in numerical taxonomy, Euclidean distances generally are standardized (Everitt, 1974, p. 56-57), however, no problems of scale existed as the input data were all transition probabilities and constrained between zero and unity. It also was desirable that the lithologic transitions with the greatest variance should exert a greater influence on the final classification than transitions that ranged little from section to section (Sneath and Sokal, 1973, p. 105). Raw, as opposed to standardized, Euclidean distances therefore were calculated from the transition probability vectors.

The problem of selecting the most suitable classification technique was more difficult than selecting the most suitable similarity coefficient because of the multiplicity of techniques available. Cluster analysis was desirable to use because it had been employed in the Read and Sackin (1971) and Read and Merriam (1972) studies, and the dendrograms would provide an easy, visual means of comparing the results. Single-linkage cluster analysis has the advantages of simplicity and mathematical acceptability but is subject to the "chaining" effects if intermediate points are present between the main clusters (Sokal and Sneath, 1963, p. 190; Sneath and Sokal, 1973, p. 218; Everitt, 1974, p. 61, 93). The nature of the basic data, namely a succession of partially overlapping delta-lobes, makes it highly probable that such intermediate points will exist. This, combined with the unsatisfactory classifications produced by single-linkage cluster analysis in earlier studies (Read and Sackin, 1971, fig. 3; Read and Merriam, 1972, fig. 1), led to its rejection in favor of average-linkage methods.

The average-linkage method employed by Read and Merriam (1972) to cluster the correlation coefficient data was the Rohlf (1963) unweighted arithmetic method. However, in view of the comments of Sokal and Sneath (1963, p. 184, 310), it was decided in this study to cluster the correlation coefficients using the average-linkage algorithm given in the ROKDOC package (Loudon, 1967; 1974, p.17). The Euclidean distances were clustered using Rohlf's (1963) unweighted arithmetic method, which calculates simple averages. The ROKDOC package's algorithm recomputes the correlation coefficients at the end of each clustering cycle using Spearman's sums of variables method and reverals of correlation level are possible (Fig. 2; Sokal and Sneath, 1963, p. 183-184), whereas such reversals cannot occur in Rohlf's (1963) unweighted arithmetic method.

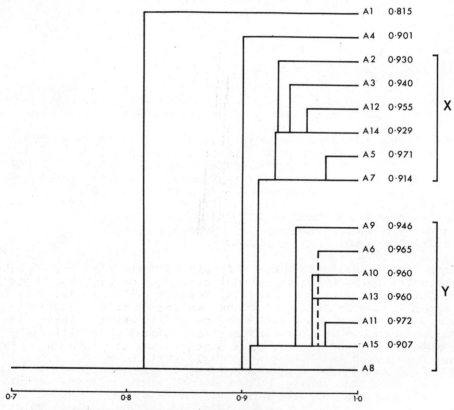

Figure 2. Dendrogram showing average-linkage cluster analysis (ROKDOC package) of correlation coefficients listed by Read and Dean (1972, table 3).

Because of the likelihood of intermediate points, it was necessary to check the results of cluster analysis by classifying the data by some sort of ordination method. Instead of using Q-mode factor analysis, the method of nonlinear mapping developed by Sammon (1969) and applied to geological data by Henley (1972) and Howarth (1973) was selected because the results are intrinsically simpler to interpret. The method can be used effectively to demonstrate the relationships between the sections in only two dimensions (Everitt, 1974, p. 94-95). The transition-probability data listed by Read and Merriam (1972, table 2) were used as input for Henley's (1972) program. This calculated the Euclidean distances between the 15 sections in 17-dimensional variable-space and progressively reduced the error indicating how well the structure of the 15 points in two-dimensional space reproduced their structure in the original variable-space (see Henley, 1972, p. 9-10).

Finally, an approximate idea of the relation between the various similarity coefficients and cophenetic values shown by the dendrograms and by the nonlinear mapping plot, was determined by calculating matrix coefficients (Sneath and Sokal, 1973, p. 278-280) for every possible pair of matrices listed here:

S1  Crossassociation similarity coefficients (see Read and Sackin, 1971, table 2).
S2  Correlation coefficients derived from vectors of transition probabilities (see Read and Merriam, 1972, table 3).
S3  Euclidean distances derived from vectors of transition probabilities (see Table 2).
C1  Cophenetic values given by dendrogram for average-linkage cluster analysis (ROKDOC package) of S2 values (see Fig. 2).
C2  Cophenetic values given by dendrogram of unweighted arithmetic cluster analysis (Rohlf, 1963) of S3 values (see Fig. 3).
C3  Cophenetic values (distances between sections) from the final two-dimensional plot produced by Henley's (1962) nonlinear mapping program.

In order to avoid confusion, it should be noted that the numbers used to designate the matrices do not correspond to those used by Read and Sackin (1971). Read and Merriam (1972, p. 224-225) suggested the possibility of either weighting selected transitions, which were judged either to be of low environmental significance or else to have similar transitional probabilities for all sections. This has not been done because of the problems inherent in weighting (Sneath and Sokal, 1973, p. 6, 14, 109-113) and the dangers of introducing subjective judgement.

## RESULTS

The results of the overall comparison of the 15 strings of lithologic states using the crossassociation similarity coefficient are given in detail in Read and Sackin (1971, p. 10-11, fig. 2, table 2). The average-linkage dendrogram divided the

Table 2. Matrix of Euclidean distances (Matrix S3) derived from transition probability data listed by Read and Merriam (1972, table 2)

|  | A1 | A2 | A3 | A4 | A5 | A6 | A7 | A8 | A9 | A10 | A11 | A12 | A13 | A14 | A15 |
|---|---|---|---|---|---|---|---|---|---|---|---|---|---|---|---|
| A1 | 0.00 | | | | | | | | | | | | | | |
| A2 | 0.84 | 0.00 | | | | | | | | | | | | | |
| A3 | 0.75 | 0.71 | 0.00 | | | | | | | | | | | | |
| A4 | 0.71 | 0.63 | 0.68 | 0.00 | | | | | | | | | | | |
| A5 | 0.84 | 0.57 | 0.70 | 0.40 | 0.00 | | | | | | | | | | |
| A6 | 0.93 | 0.47 | 0.62 | 0.62 | 0.38 | 0.00 | | | | | | | | | |
| A7 | 0.82 | 0.66 | 0.64 | 0.47 | 0.21 | 0.45 | 0.00 | | | | | | | | |
| A8 | 1.25 | 0.87 | 1.01 | 0.86 | 0.60 | 0.52 | 0.66 | 0.00 | | | | | | | |
| A9 | 0.88 | 0.60 | 0.58 | 0.63 | 0.44 | 0.28 | 0.42 | 0.53 | 0.00 | | | | | | |
| A10 | 1.22 | 0.80 | 0.74 | 0.92 | 0.64 | 0.46 | 0.66 | 0.65 | 0.52 | 0.00 | | | | | |
| A11 | 1.19 | 0.66 | 0.80 | 0.86 | 0.64 | 0.38 | 0.68 | 0.50 | 0.39 | 0.42 | 0.00 | | | | |
| A12 | 0.78 | 0.57 | 0.46 | 0.50 | 0.41 | 0.38 | 0.42 | 0.75 | 0.42 | 0.52 | 0.60 | 0.00 | | | |
| A13 | 1.11 | 0.59 | 0.77 | 0.65 | 0.43 | 0.31 | 0.52 | 0.50 | 0.42 | 0.41 | 0.35 | 0.43 | 0.00 | | |
| A14 | 0.72 | 0.45 | 0.66 | 0.43 | 0.34 | 0.36 | 0.40 | 0.71 | 0.45 | 0.72 | 0.66 | 0.38 | 0.49 | 0.00 | |
| A15 | 1.08 | 0.54 | 0.79 | 0.70 | 0.47 | 0.30 | 0.55 | 0.38 | 0.37 | 0.51 | 0.29 | 0.53 | 0.27 | 0.50 | 0.00 |

sections into three clusters. The first comprised A1 alone, the second comprised A2-4, A7, A8 and A10, and the third the remainder of the sections (Read and Sackin, 1971, fig. 2). The cophenetic correlation coefficient value for this dendrogram was only 0.50, which is unusually low (Sneath and Sokal, 1973, p. 278).

In the unweighted arithmetic clustering of correlation coefficient data by Read and Merriam (1972, fig. 2) the sections again were divided into three clusters which showed some basic similarities to the groups obtained by the crossassociation study. A1 again formed a cluster by itself. The second cluster, which was termed Cluster X, again contained A2-4 and A7, but also it contained A5, A12 and A14 and did not contain A8 and A10. The third cluster, termed Cluster Y, contained A6, A8-11, A13 and A15. A1 shows a high probability for the mudstone to sandstone transition. In the sections in Cluster X this transition probability was greater than, or about equal to, that for mudstone to siltstone, whereas in the sections in Cluster Y the former transition probability was lower than the latter (Read and Merriam, 1972, table 2).

In the new variants of this method the grouping remained similar, which indicated that the basic method was reasonably robust (Everitt, 1974, p. 66-67). When the correlation coefficients (Read and Merriam, 1972, table 3) were clustered using the ROKDOC algorithm (Fig. 2), A4 remained independent of Cluster X. Similar to A1, this section has a high probability for the mudstone to sandstone transition and a low probability for the mudstone to siltstone transition. When the Euclidean distances (Table 2) were clustered using Rohlf's (1963) unweighted arithmetic method (Fig. 3), A4 joined Cluster X at a late stage but A3 remained independent. This section has almost as high a probability for the mudstone to sandstone transition as A4 but, in addition, it has a particularly high probability for the siltstone to sandstone transition.

# QUANTITATIVE METHODS OF COMPARING SUCCESSION DATA 41

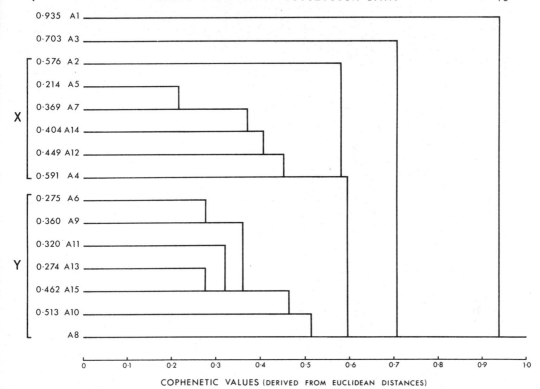

Figure 3. Dendrogram showing unweighted arithmetic cluster analysis (Rohlf, 1963) of Euclidean distances listed in Table 2.

The most interesting results, however, came from the two-dimensional plot produced by Henley's (1972) nonlinear mapping program (Fig. 4). This showed that the sections were not naturally grouped into dense clusters but that there was a gradual transition from the sections of Cluster X to those of Cluster Y, with A5, A7 and A12 of the former cluster and A6 and A9 of the latter cluster occupying transitional positions. Significantly, the transition probabilities for mudstone to sandstone and for mudstone to siltstone are equal in A5, A7 and A12. The plot also emphasized the relative isolation not only of A1 but also of A2, and A3 and A4. A2, unlike the other sections in Cluster X, has a probability for the mudstone to sandstone transition which is slightly lower than that for the mudstone to siltstone transition and shows high probabilities for the sandstone to seatrock and coal to seatrock transitions. In general the spacing of the 15 sections along Axis I, the mantissa, is linked closely to their transition probabilities for mudstone to sandstone and for mudstone to siltstone.

The values of the matrix coefficients used to compare similarity coefficient values $r_{(S1, S2)}$, $r_{(S1,S3)}$ and $r_{(S2, S3)}$ (Table 3) reveal that the correlation coefficient values and the Euclidean distances are strongly negatively correlated and that there is some degree of

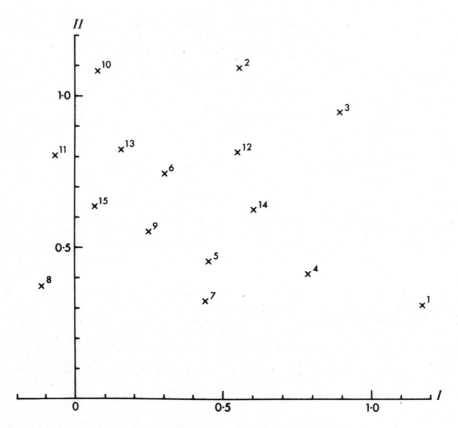

Figure 4. Two-dimensional plot produced by Henley's (1972) nonlinear mapping program from Euclidean distances listed in Table 2.

correlation between both sets of values and those of the cross-association similarity coefficients. The cophenetic correlation coefficient values $r_{(S2, C1)}$, $r_{(S3, C2)}$ and $r_{(S3, C3)}$ show that the dendrogram derived from the average-linkage cluster analysis (ROKDOC program) of the correlation coefficient is less effective in representing the input data than the dendrogram derived from the unweighted arithmetic cluster analysis (Rohlf, 1963) of the Euclidean distances. The latter dendrogram is, however, itself a less effective method of representing the original Euclidean distances than is the two-dimensional plot produced by Henley's (1972) nonlinear mapping program. The high value for $r_{(C1, C2)}$ reveals the close similarity between the two dendrograms mentioned.

## DISCUSSION AND SEDIMENTOLOGIC INTERPRETATION

The results as outlined confirm the impression that cross-association is not a particularly effective technique for the overall comparison of lithologic sequences. This is mainly due

Table 3. Matrix coefficients derived from comparing all possible pairs of matrices of similarity coefficients (Matrices S1-3) and cophenetic values (Matrices C1-3).

|    | S1    | S2    | S3    | C1    | C2    | C3   |
|----|-------|-------|-------|-------|-------|------|
| S1 | 1.00  |       |       |       |       |      |
| S2 | 0.59  | 1.00  |       |       |       |      |
| S3 | -0.57 | -0.91 | 1.00  |       |       |      |
| C1 | 0.53  | 0.70  | -0.74 | 1.00  |       |      |
| C2 | -0.49 | -0.66 | 0.81  | -0.89 | 1.00  |      |
| C3 | -0.57 | -0.85 | 0.93  | -0.66 | 0.72  | 1.00 |

S1 = Matrix of cross-association similarity coefficient values

S2 = Matrix of correlation coefficient values derived from vectors of transition probabilities

S3 = Matrix of Euclidean distances derived from vectors of transition probabilities

C1 = Matrix of cophenetic values given by the dendrogram for average linkage cluster analysis (ROKDOC package) of S2 values

C2 = Matrix of cophenetic values given by the dendrogram for unweighted arithmetic cluster analysis (Rohlf 1963) of S3 values

C3 = Matrix of cophenetic values (final distances) produced by the nonlinear mapping (Henley 1972) of S3 values.

to the method of calculating the crossassociation similarity coefficient, which is designed to measure the properties of the two sequences that can be paired as matching subsequences. In practice, with this data set which has a definite cyclical structure, short identical subsequences at the bottom of one sequence and the top of the other are matched and deleted. This occurs before the longer identical subsequences of which they form a part can come together in later matching positions. Most of the subsequences matched and deleted are in fact short and, in addition, many are simple alternations of siltstone and sandstone or seatrock and coal which have little overall sedimentolgic significance. The method also suffers from the defect that it provides no details of the actual structure of the data or of the most commonly occurring subsequences, yet this is precisely the type of information required to determine the environment of deposition. Thus crossassociation can indicate that the strings of lithologies possess a cyclic structure but cannot give details of the type of cycle present.

In view of these shortcomings, it is interesting to note that some degree of correlation exists firstly between the crossassociation similarity coefficients and the other two sets of similarity coefficients and secondly between the dendrograms produced by clustering these three sets of similarity coefficients. Thus it was possible for Read and Sackin (1971, p. 13) to deduce that A1 had a more or less unique depositional environment and that A2-4 and A7 shared a similar depositional environment.

The technique developed by Read and Merriam (1972) revealed just what these environments were and the new variants on this basic technique described in this paper aid the sedimentologic interpretation further. Thus A1 lies at a point where erosive channels carrying coarse-grained clastic sediments entered the Kincardine Basin from the northwest. The Limestone Coal Group sediments of the Kincardine Basin closely resemble modern deltaic deposits (Read, 1965, 1969a, 1969b, 1970) and are thought to represent a series of overlapping high-constructive delta-lobes, each of which was built out, colonized by vegetation, and finally abandoned (Scruton, 1960; Coleman and Cagliano, 1964; Frazier, 1967; Fisher, 1969). The high probability for the transition from mudstone up to sandstone in A1 represents the high probability of encountering an erosive distributary channel which has cut through an earlier sequence of upward-coarsening distributary mouth-bar sands and eventually has been filled with an upward-fining sequence of sands (Fisk, 1955, 1960; Fisher and others, 1969). Upward-fining cycles, which are characteristic of channel-fill sequences (Reading, 1967) are dominant in A1.

In the Limestone Coal Group of the Kincardine Basin, two belts which were commonly occupied by such erosive channels have been recognized (Read, 1970). One, which is here termed the Stirling Channel Belt, enters the northwestern part of the basin and extends southwards whereas the other, which is termed the Saline Channel Belt, enters the northeastern part of the basin and follows a curving course southwards and westwards to join the Stirling Channel Belt in the southwest of the basin. Evidence from the sandstone/shale ratios in the upper part of the Limestone Coal Group (Read and Dean, 1967, fig. 4; 1970) and from detailed borehole sections has been used to plot the center lines of the channel belts on Figure 1 which shows that A1 lies near the middle of the Stirling Channel Belt.

The relatively high probabilities for the mudstone to sandstone transition in the sections in Cluster X reflects the frequent presence of these erosive channels filled with upward-fining sands, although these channels are less common than in A1. Figure 1 shows that most of the sections of the sections in Cluster X lie fairly close to the centerlines of the channel belts. There are however two exceptions to this rule. The first is A5 which belongs to Cluster X in Figures 2 and 3 (see also Read and Merriam, 1972, fig. 2) but does not lie close to either channel belt. The second is A9 which belongs to Cluster Y yet lies close to the Saline Channel Belt. The nonlinear mapping plot (Fig. 4) indicates the intermediate classification positions of these two sections and it is probably significant that a particularly important channel complex is known to lie in the lower part of the succession studied at the site of A5 (Read, 1961, fig. 8). A3 and A4 are characterized by a higher probability for mudstone to sandstone transition than are the other sections in Group X. This reflects the more common occurrence of erosive channels and it is not surprising to see in Figure 1 that both sections lie close to the center of the combined channel belt. Thus the "unstable" positions of A3 and A4 in Figures 2 and 3 and the relatively isolated positions of these sections on Figure 4 indicate a definite difference in depositional environment. The somewhat isolated position of A2 in Figures 2-4 partly reflects the high probabilities for the sandstone to seatrock and coal to seatrock transitions. Both probably reflect frequent floods bringing layers of clastic sediment into a marsh or swamp environment that lay close to a channel belt.

The remaining sections, which are grouped together in Cluster Y, all have transition probabilities for mudstone and sandstone which are lower than those for mudstone to siltstone. This demonstrates the dominance of the upward-coarsening cycles that are typical of most deltaic successions (Reading, 1967). These are best developed in the interdistributary trough areas remote from the major distributaries, where prodelta clays pass by alternation gradually upwards into upward-coarsening delta-front sheet sands and silty sands which are redistributed from distributary mouth bars by waves and currents (Fisk, 1955; Fisher and McGowen, 1969; Gould, 1970; Guevara and Garcia, 1972).

Predictably, the sections in Cluster Y generally tend to lie towards the center of the basin in areas fairly remote from the channel belts (Fig. 1).

The dendrograms shown in Figures 2 and 3 are at least as easy to interpret in sedimentologic terms as the equivalent dendrogram in the Read and Merriam (1972, fig. 2) study and, especially if considered together, they emphasize new sedimentolgic aspects of the data set. All the dendrograms by their nature however force the sections into clusters (Sneath and Sokal, 1973, p. 252). The true relations between the sections are more clearly indicated by the nonlinear mapping plot (Fig. 4) which is consequently more readily interpretable from a sedimentologic standpoint than any of the dendrograms. In particular it brings out the lack of any really sharp division into well-defined groups, which is what would be expected in a depositional environment characterized by a series of overlapping delta-lobes. A study of the detailed stratigraphy reveals that the channel belts migrated extensively during upper Limestone Coal Group time. Thus the boundaries between areas characterized by upward-fining cyclothems with channel sandstones and areas characterized by upward-coarsening cyclothems with sheet sandstones would be expected to be blurred. This evidence, together with that of the matrix coefficients, suggests that of all the techniques investigated in this study, that of classifying Euclidean distances by means of nonlinear mapping is the most effective.

## POSSIBLE FUTURE DEVELOPMENTS

The method of calculating similarity coefficients between vectors of transition probabilities suffers from the disadvantage that each row of the transition-probability matrix from which these vectors are derived is constrained to sum to unity. The transition-probability vector is a series of closed sets and this can lead to strong negative correlation between particular transition probabilities (Read and Merriam, 1972, p. 210), which can, in turn, lead to a form of weighting. In this data set, the transition probabilities for mudstone to sandstone and for mudstone to siltstone are known to be negatively correlated but fortunately these transitions reflect the most important sedimentologic feature in the succession, namely the relative abundance of upward-fining and upward-coarsening cycles.

An alternative approach would be to discard all transitions with probabilities that are more or less the same in all the sections because these transitions could be regarded as invariant characters (Sneath and Sokal, 1973, p. 105) and to select only one transition from each closed set. If the conceptual depositional model is known, then the selection can be made on the grounds of sedimentologic significance but if not, the transition with the maximum variance could be selected. Other similarity coefficients and other classification methods also could be tried.

## CONCLUSIONS

More work needs to be done on techniques for the quantitative overall comparison of lithologic successions before mathematical basin analysis can develop its full potential.

Crossassociation, which has been the most widely used technique to date, suffers from the defects that the subsequences which are pared are generally short, that undue weight is given to simple alternations of low sedimentologic significance and that the technique provides comparatively little insight into the sedimentologic processes involved. It has proved to be only partially effective in analyzing this data set.

The calculation of similarity coefficients between vectors of transitional probabilities has provided a better insight into the depositional processes and the results are more readily interpretable from a sedimentologic standpoint. The average-linkage cluster analysis (ROKDOC program) of correlation coefficients and the unweighted arithmetic cluster analysis (Rohlf, 1963) of Euclidean distances have revealed features of sedimentologic interest which were not brought out by the original Read and Merriam (1972) technique. Nevertheless, the dendrograms force all the sections into clusters and do not reveal either the absence of any well-defined, dense natural clusters or the intermediate nature of many of the sections. Henley's (1972) nonlinear mapping program, which uses Euclidean distances, does however emphasize these features and also reveals the complex relationships between the sections more clearly. In particular this method reveals the relative importance of upward-fining cycles, which contain erosive sand-filled channels, and upward-coarsening cycles, which contain delta-front sheet sands, in the various sections. Confirmatory evidence is provided by the geographic position of the section relative to known channel belts. Of the techniques investigated, the calculation of Euclidean distances from vectors of transition probabilities and the classification of the results by nonlinear mapping is the one which is judge to be most effective.

## ACKNOWLEDGMENTS

The author is particularly grateful firstly to Dr. S. Henley of the IGS Computer Unit in Edinburgh for pointing out the potential usefulness of nonlinear mapping in the context of this problem and for processing the data using his (1972) program at the Edinburgh Regional Computer Centre. Secondly to Mrs. Elizabeth Gill for adapting the ROKDOC cluster analysis package to use the Rohlf (1963) algorithm and doing all the other data-processing involved in this study at the SRC Atlas Computer Laboratory. Without their kind assistance it would have been impossible to complete the study. The author also wishes to thank the following for reading the paper and making helpful suggestions for its improvement: Prof. E.S. Belt, Dr. K.G. Jeffery, Dr. T.V. Loudon, Prof. D.F. Merriam, Dr. P.A. Sabine, and Dr. R.C. Selley.

## REFERENCES

Allègre, C., 1964, Vers une logique mathematique des series sedimentaires: Geol. Soc. France Bull., Ser. 7, tome 6, no. 2, p. 214-218.

Allen, P., 1959, The Wealden environment: Anglo-Paris basin: Phil. Trans. Royal Soc. London, Ser. B., v. 242, p. 283-346.

Allen, P., and Krumbein, W.C., 1962, Secondary trend components in the Top Ashdown Pebble Bed: a case history: Jour. Geology, v. 70, no. 5, p. 507-538.

Carr, D.D., Horowitz, A., Hrabar, S.V., Ridge, K.F., Rooney, R., Straw, W.R., Webb, W., and Potter, P.E., 1966, Stratigraphic processes, bedding sequences and random processes: Science, v. 154, no. 3753, p. 1162-1164.

Coleman, J.M., and Gagliano, S.M., 1964, Cyclic sedimentation in the Mississippi River deltaic plain: Gulf - Coast Assoc. Geol. Socs. Trans., v. 14, p. 67-80.

Dean, W.E., and Anderson, R.Y., 1974, Application of some correlation coefficient techniques to time-series analysis: Jour. Math. Geology, v. 6, no. 4, p. 363-372.

de Raaf, J.M.F., Reading, H.G., and Walker, R.G., 1965, Cyclic sedimentation in the Lower Westphalian of North Devon, England: Sedimentology, v. 4, no. 1/2, p. 1-52.

Duff, P. McL.D., and Walton, E.K., 1964, Trend surface analysis of sedimentary features of the Modiolaris Zone, east Pennine Coalfield, England, in Deltaic and shallow marine deposits: Elsevier Publ. Co., Amsterdam, p. 114-122.

Everitt, B., 1974, Cluster analysis: Heinemann, London, 122 p.

Fisk, H.N., 1955, Sand facies of recent Mississippi delta deposits: Proc. 4th World Petroleum Congress, Sect. 1/C, Paper 3, p. 377-398.

Fisk, H.N., 1960, Recent Mississippi River sedimentation and peat accumulation: Congrès Advan. Etud. Stratigr. Carb., p. 187-199.

Fisher, W.L., 1969, Facies characterization of Gulf Coast Basin delta systems with some Holocene analogues: Gulf-Coast Assoc. Geol. Socs. Trans., v. 19, p. 239-261.

Fisher, W.L., Brown, L.F., Scott, A.J., and McGowen, J.H., 1969, Delta systems in the exploration for oil and gas: a research colloquium: Bur. Econ. Geology, Univ. Texas at Austin, 78 p.

Fisher, W.L., and McGowen, J.H., 1967, Depositional systems in the Wilcox Group of Texas and their relationship to occurrence of oil and gas: Gulf-Coast Assoc. Geol. Socs. Trans., v. 17, p. 105-125.

Fisher, W.L., and McGowen, J.H., 1969, Depositional systems in the Wilcox Group (Eocene) of Texas and their relationship to occurrence of oil and gas: Am. Assoc. Petroleum Geologists Bull., v. 53, no. 1, p. 30-54.

Fox, W.T., and Brown, J.A., 1965, The use of time-trend analysis for environmental interpretation of limestones: Jour. Geology, v. 73, no. 3, p. 510-518.

Frazier, D.E., 1967, Recent deltaic deposits of the Mississippi River: their development and chronology: Gulf-Coast Assoc. Geol. Socs. Trans., v. 17, p. 287-315.

Friend, P.F., and Moody-Stuart, M., 1972, Sedimentation of the Wood Bay Formation (Devonian) of Spitzbergen: regional analysis of a late orogenic basin: Norsk Polarinstitutt Skrifter, No. 157, 77 p.

Goodlet, G.A., 1957, Lithological variation in the Lower Limestone Group of the Midland Valley of Scotland: Geol. Survey Great Britain Bull., No. 12, p. 52-65.

Gould, H.R., 1970, The Mississippi delta complex, in Deltaic sedimentation modern and ancient: Soc. Econ. Paleont. Mineralogy, Sp. Publ. 15, p. 3-30.

Guevara, E.H., and Garcia, R., 1972, Depositional systems and oil-gas reservoirs in the Queen City Formation (Eocene), Texas: Gulf-Coast Assoc. Geol. Socs. Trans., v. 12, p. 1-22.

Henley, S., 1972, Nonlinear mapping and a related R-mode technique for compression of multivariate data: Bur. Mineral Resources, Geology and Geophysics Record No. 1972/124, 12 p.

Howarth, R.J., 1973, Preliminary assessment of a nonlinear mapping algorithm in a geological context: Jour. Math. Geology, v. 5, no. 1, p. 39-57.

Johnson, K.R., and Cook. A.C., 1973, Cyclic characteristics of sediments in the Moon Island Beach Subgroup, Newcastle Coal Measures, New South Wales: Jour. Math. Geology, v. 5, no. 1, p. 91-110.

Kelling, G., 1968, Patterns of sedimentation in Rhondda Beds of South Wales: Am. Assoc. Petroleum Geologists Bull., v. 52, no. 12, p. 2369-2386.

Krumbein, W.C., 1948, Lithofacies maps and regional sedimentary stratigraphic analysis: Am. Assoc. Petroleum Geologists Bull., v. 32, no. 10, p. 1909-1923.

Krumbein, W.C., 1967, FORTRAN IV computer program for Markov chain experiments in geology: Kansas Geol. Survey Computer Contr. 13, 38 p.

Loudon, T.V., 1967, The ROKDOC package: Sed. Res. Lab., Reading Univ., Rept. No. 5, 151 p.

Loudon, T.V., 174, Analysis of geological data using ROKDOC, a FORTRAN IV package for the IBM 360/65 computer: Inst. Geol. Sci., Rept. No. 74/1, 130 p.

Merriam, D.F., and Sneath, P.H.A., 1967, Comparison of cyclic rock sequences using cross-association, *in* Essays in paleontology and stratigraphy: Univ. Kansas Dept. Geology, Sp. Publ. 2, p. 521-538.

Potter, P.E., and Blakely, R.F., 1967, Generation of synthetic vertical profile of a fluvial sandstone body: Jour. Soc. Petroleum Geology, v. 7, no. 3, p. 243-251.

Potter, P.E., and Blakely, R.F., 1968, Random processes and lithological transitions: Jour. Geology, v. 76, no. 2, p. 154-170.

Potter, P.E., and Pettijohn, F.J., 1963, Palaeocurrents and basin analysis: Springer-Verlag, Berlin, 296 p.

Read, W.A., 1959, The economic geology of the Stirling and Clackmannan Coalfield, Scotland. Area south of the River Forth: Coalfield. Paper Geol. Survey, No. 2, 73 p.

Read, W.A., 1961, Aberrant cyclic sedimentation in the Limestone Coal Group of the Stirling Coalfield: Edinburgh Geol. Soc. Trans., v. 18, p. 271-292.

Read, W.A., 1965, Shoreward facies changes and their relation to cyclical sedimentation in part of the Namurian east of Stirling, Scotland: Scottish Jour. Geology, v. 1, pt. 1, p. 69-92.

Read, W.A., 1969a, Fluviatile deposits in Namurian rocks of central Scotland: Geol. Mag., v. 106, p. 331-347.

Read, W.A., 1969b, Analysis and simulation of Namurian sediments in central Scotland using a Markov-process model: Jour. Math. Geology, v. 1, no. 2, p. 199-219.

Read, W.A., 1970, Cyclically-deposited Namurian sediments east of Stirling, Scotland: unpubl. doctoral dissertation, Univ. London, 273 p.

Read, W.A., and Dean, J.M., 1967, A quantitative study of a sequence of coal-bearing cycles in the Namurian of central Scotland, 1: Sedimentology, v. 9, no. 2, p. 137-156.

Read, W.A., and Dean, J.M., 1968, A quantitative study of a sequence of coal-bearing cycles in the Namurian of central Scotland, 2: Sedimentology, v. 10, no. 2, p. 121-136.

Read, W.A., and Dean, J.M., 1972, Principal component analysis of lithological variables from some Namurian ($E_2$) paralic sediments in central Scotland: Geol. Surv. Great Britain Bull., No. 40, p. 83-99.

Read, W.A., Dean, J.H., and Cole, A.J., 1971, Some Namurian ($E_2$) paraplic sediments in central Scotland: an investigation of depositional environment and facies changes using iterative-fit trend-surface analysis: Geol. Soc. London Quart. Jour. v. 127, pt. 2, p. 137-176.

Read, W.A., and Merriam, D.F., 1972, A simple quantitative technique for comparing cyclically deposited successions, in Mathematical models of sedimentary processes: Plenum Press, New York, p. 203-231.

Read, W.A., and Sackin, M.J., 1971, A quantitative comparison, using cross-association of vertical sections of Namurian ($E_1$) paralic sediments in the Kincardine Basin, Scotland: Inst. Geol. Sci., Rept. No. 71/14, 21 p.

Reading, H.G., 1967, Sedimentation sequences in the Upper Carboniferous of northwestern Europe: Congrès Advan. Etud Stratigr. Carb., v. 4, p. 1401-1412.

Rohlf, F.J., 1963, Classification of *Aedes* by numerical taxonomic methods (Diptera; Culcidae): Ann. Entomol. Soc. America, v. 56, p. 798-804.

Sackin, M.J., and Sneath, P.H.A., 1965, Amino acid sequences in proteins: a computer study: Biochem. Jour. v. 96, p. 70p-71p.

Sackin, M.J., Sneath, P.H.A., and Merriam, D.F., 1965, ALGOL program for cross-association of nonnumeric sequences using a medium-sized computer: Kansas Geol. Survey Sp. Dist. Publ. 23, 36 p.

Sammon, J.W., 1969, A nonlinear mapping for data structure analysis: IEEE Trans. Computers C-18, no. 5, p. 401-409.

Schwarzacher, W., 1967, Some experiments to simulate the Pennsylvanian rock sequence of Kansas: Kansas Geol. Survey Computer Contr. 18, p. 5-14.

Schwarzacher, W., 1969, The use of Markov chains in the study of sedimentary cycles: Jour. Math. Geology, v. 1, no. 1, p. 17-39.

Scruton, P.C., 1960, Delta building and the deltaic sequence, *in* Recent sediments, northwest Gulf of Mexico: Am. Assoc. Petroleum Geologists, Tulsa, p. 82-102.

Selley, R.C., 1970, Studies of sequence in sediments using a simple mathematical device: Geol. Soc. London Quart. Jour., v. 125, p. 577-581.

Sneath, P.H.A., and Sokal, R.R., 1973, Numerical taxonomy: W.H. Freeman and Co., San Francisco, 573 p.

Sokal, R.R., and Sneath, P.H.A., 1963, Principles of numerical taxonomy: W.H. Freeman and Co., San Francisco, 359 p.

Visher, G.S., 1965, Uses of vertical profile in environmental reconstruction: Am. Assoc. Petroleum Geologists Bull., v. 49, no. 1, p. 41-61.

Vistelius, A.B., 1949, On the question of the mechanism of the formation of strata: Dokl. Akad. Nauk SSSR, v. 65, no. 2, p. 191-194.

Vistelius, A.B., 1961, Sedimentation time trend functions and their application for correlation of sedimentary deposits: Jour. Geology, v. 69, no. 6, p. 703-728.

# STATISTICAL RECOGNITION OF TERRESTRIAL AND MARINE SEDIMENTS IN THE LOWER CRETACEOUS OF PORTUGAL

R.A. Reyment, P.Y. Berthou, and B.Å. Moberg

*Uppsala University, University of Paris, and Uppsala University*

## ABSTRACT

The environmental nature of Lower Cretaceous rocks of the coastal cliff section of Praia da Luz in the Algarve Province, Portugal, was studied by multivariate statistical analysis of geochemical determinations. The paleontologically based assignments to a terrestrial or marine origin were confirmed mainly by the statistical work. KEY WORDS: *graphics, factor analysis, stratigraphy*.

## INTRODUCTION

A marked coastal feature of the western Algarve in the vicinity of the city of Lagos, between Praia da Luz and Porto de Mos, is a line of steep cliffs, composed mainly of sedimentary rocks of Early Cretaceous age. Near Punta da Piedade, Miocene deposits lie unconformably on the Cretaceous beds.

Rey, Grambast, and Ramalho (1974) gave a description of the Mesozoic sequence along the Algarvian coast, from the village of Burgau, 10 km west of Lagos, to Atalaia. These deposits represent several Lower Cretaceous stages and range from Berriasian to Upper Aptian ("Gargasian") with a probable gap in the Lower Hauterivian. This sequence extends to the top of the marls at Praia da Luz, forming the cliff beneath the trig-point at Atalaia. According to these workers, the upper part of the marls is Late Aptian in age. Above the marls, toward the village of Porto de Mos, there is an alternating sequence of argillaceous limestones, marls, and shales, which forms the continuation of the sequence into the Tertiary of Punta da Piedade. These beds are poor in fossils and the microfauna is insufficient to allow of exact dating. Albian deposits are not known in to occur with certainty and Middle and Upper Cretaceous sediments would seem to be missing in the area.

Our samples were collected by Berthou and Reyment in April 1973, from the Praia da Luz section, which corresponds to level 7 of Rey, Grambast, and Ramalho (1974). The marls of the sequence, which may be rich in charophytes, seem to correspond to a margino-litoral sedimentary environment. There are numerous alternations between marine and nonmarine beds throughout the section.

The aim of our study has been to ascertain whether the environmental shifts indicated by the macroscopic appearance of the rocks and the study of thin sections can be confirmed by means of geochemical determinations on the sediments, analyzed by suitable statistical methods.

## SAMPLING DETAILS

Samples were taken from 17 clearly defined beds in the Praia da Luz section. In many situations, the macroscopical features of the rocks are sufficient to disclose the environment of deposition of the original sediment. For example, at the base of the cliffs, there are algal mats in the calcareous sandstones, worm-tubes, cross-bedded sandstones and and sandstones with abundant plant fragments.

## METHODS

The field observations were supplemented by a study of thin sections where possible. The two sets of data were employed in conjunction to form the basis of decision as to whether a particular rock was marine or nonmarine in origin. The geochemical determinations were analyzed by standard methods of multivariate statistics to determine whether these decisions could be duplicated. The results of the qualitatively made decisions are listed in Table 1. It will be seen from this table that all sediments seem to be of strand-near origin, thus implying that the history of the area during the time of deposition of the strata was marked by slight oscillations in sealevel.

Table 1. Result of study of thin sections of sedimentary rock specimens

| Specimen | Description | Assignment |
|---|---|---|
| 1 | Gray, fine-grained, secondarily calcified sandstone: loosely consolidated, with quartz, biotite, orthoclase. No fossils. | terrestrial |
| 2 | Pale-yellow to white, compact limestone, rich in shell fragments, including calcareous algae, pelecypods and foraminifers. Some recrystallized calcite. | marine |

| Specimen | Description | Assignment |
|---|---|---|
| 3 | Sandstone of the same type as specimen 1. | terrestrial |
| 4 | Gray, fine-grained calcareous (secondary) sandstone containing large quartz pieces and feldspar. | terrestrial |
| 5 | Gray, unconsolidated shale, pyritized in places. No fossils. | terrestrial |
| 6 | Greenish, compact limestone with reddish traces: rich in shell fragments and well preserved pelecypods. | marine |
| 7 | Green, fine-grained sandstone containing quartz and feldspar; outcrop displays current bedding. No fossils. | terrestrial |
| 8 | Rock same type as specimen 7. | terrestrial |
| 9 | Red-brown concretionary rock intermingled with pink grit. | terrestrial |
| 10 | Gray-green, compact argillaceous limestone, rich in fossils, particulary ostracods and pelecypods. Contains small pyritic nodules. | marine |
| 11 | Light-gray, fine-grained pure sandstone, with charophytes. Large quartz pebbles visible in outcrop. | terrestrial? |
| 12 | Gray, slightly argillaceous limestone with shell fragments of mainly gastropods. Well-rounded quartz grains. | marine |
| 13 | Gray, fine-grained marl, rich in fragments of pelecypods. | marine |
| 14 (2) | Green, fine-grained calcareous shale containing pelecypods, algae, and some charophytes. | marine |
| 15 | Green, fine-grained, slightly calcareous clay with abundant feldspar crystals. Contains pelecypods. | marine |
| 16 | Light-gray, fine-grained, slightly calcareous sandstone containing quartz and feldspar pieces and flakes of mica. Charophytes. | terrestrial |
| 17 | Gray, homogeneous limestone containing gastropods, pelecypods and ostracods. Forms top of cliff sequence. | marine |

## CHEMICAL ANALYSES

The chemical analyses were made on solutions of the rocks, using standard methods with a spectrophotometer or atomic absorption spectrophotometer. In general, it seems as though the determinations of Al may tend to be somewhat too low owing to the analytical method used. The following elements were determined: Ca, Mg, Si, Na, K, Al, Fe, P, Mn, Ti, Sr and V.

## STATISTICAL ANALYSIS

The geochemical determinations were analyzed statistically by Gower's method of principal coordinates, using the coefficient of similarity proposed by that author (Gower, 1971). The plot of the first and second coordinates (Fig. 1) indicates a clear subdivision of the samples into two groups. Examination of the thin sections and field observations suggests that almost all of the marine samples fall into the right-hand group, with two samples lying in the upperleft of the diagram. Specimens 13 and 15, the two which are located at a distance from the main group of points, both contain pelecypods and shell sand. The only sample that was classified visually as terrestrial in origin and which plotted with the marine samples is number 11. This seems to be of lacustrine origin in that it contains abundant charophytes belonging to several species (Rey, Grambast, and Ramalho, 1974, p. 101). However, sample 14 also contains charophytes and little doubt exists that it is of marine origin (Table 1). There is the possibility that the lacustrine environment in which the charophytes lived was immediately adjacent to the shoreline and that the charophyte fructifications were blown into the sea. Thus the environment in which the rock sample 11 was formed was probably transitional marine.

The first two principal coordinates are connected to about 44 percent of the information in the statistical sample which, for the type of data considered here, is good. Adding a third coordinate increased the information coverage by a further 14 percent. There is, therefore, good reason to study the plots of the first two coordinates with the third. The plot of the first and third coordinates repeats the result yielded by the first and second coordinates and it is clear that the first coordinate axis is responsible for ordinating the observations into terrestrial and marine deposits. Specimen 9, a concretionary rock, plots away from the main body. Although of terrestrial origin, it differs from the other specimens in that it has been secondarily altered. The impression that the third coordinate is tending to separate this weathering product is heightened by the plot of the second and third coordinates. Here, all sedimentary samples tend to group together in the lower part of the plot whereas specimen 9 occurs by itself. The principal coordinates analysis demonstrates that geochemical analyses can be used to yield an accurate reflection of the sedimentary environment in which the rocks are deposited.

Although not strictly applicable to data of the type involved

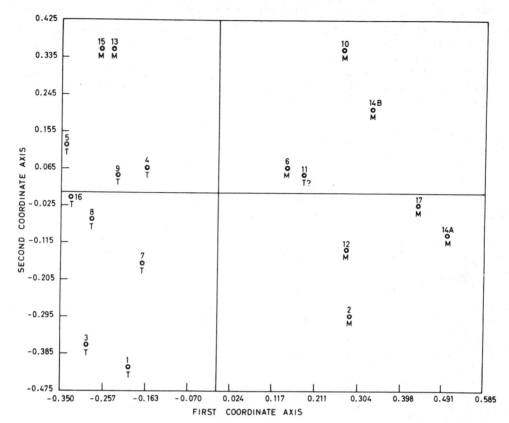

Figure 1. Plot of first and second principal coordinates by Gower's method.

here (Hill, 1974), correspondence analysis was tried tentatively on our data matrix. Correspondence analysis permits examination on the same graph of the relationship existing between objects (here, the sedimentary rock specimens) and variables (the chemical elements), which was thought to be pertinent in the present situation. A discussion of this Q-R-mode generalization of principal components analysis is given in Joreskog, Klovan, and Reyment (1975). Several combinations of plots were studied; the most instructive of these was the second and third axes (Fig. 2). Inspection of this plot discloses that the rocks identified visually as marine occur together to form a field covering most of the graph whereas the terrestrial sediments form a less widely dispersed group, with the exception of specimen 11 which, as in the coordinates analysis, seems to be out of place. The correspondence plot suggests that the terrestrial sediments are particularly influenced by the elements Si, Ti and Fe. The marine sediments, naturally of differing constitution, occur in several fields; the calcareous sediments influenced by Ca and Sr, and clastic sediments in which Mn is an important element. The role of Mg in the projection of the second and third axes is slight. Certainly there are theoretical objections to using a table of

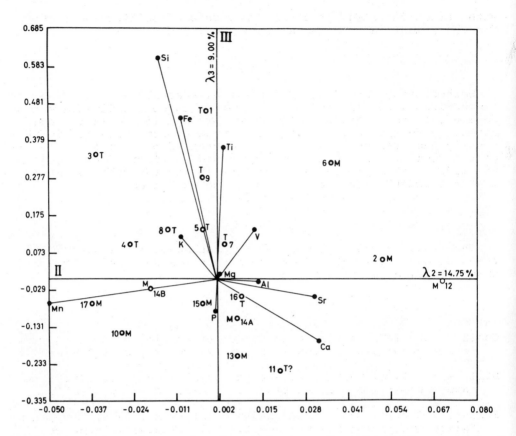

Figure 2. Plot of second and third axes of correspondence analysis

geochemical analyses as a contingency table of scores, but there is little doubt that correspondence analysis provides a reasonable interpretation and one that forms a logical complement to the co-ordinates analysis. Finally, a few remarks on the variables themselves. The correspondence projection for variables locates highly correlated characters close to each other. Figure 2 shows that Si and Fe are closely correlated as are Ca and Sr, whereas Mn is not well correlated with any of the variables.

Although the total sample is heterogeneous and composed of two environments the results of a principal factor analysis of the data matrix is of interest. Factor analysis can be accepted as a type of rough data analysis for not only is the total sample disjunct, it also is not normally distributed. The varimax rotated factor loadings, with three significant factors, encompassing 80 percent of the total variance, was employed. The first factor represents covariation in Ca, Si, Al, Mn and Sr, with Si weighted against the others. This factor seems to reflect marine sedimentation. The second factor represents covariation in Fe, P, Ti and could be interpreted as a detrital factor, reflecting the continental origin of the sediments. The third factor is concerned with variation in Mg and perhaps may be a dolomitization factor for the limestones of the material.

## SUMMARY

Methods of Q- and R-mode statistical analysis indicate that geochemical determinations on a Lower Cretaceous sedimentary rock sequence in the Algarve can be used for identifying marine and nonmarine environments. This was tested by comparing the statistically made assignments with those obtained by visual inspection of the rocks. This is a pilot study and no claims for generality are made. It is part of an investigation on the relationships in the central Atlantic during the Cretaceous Period.

## ACKNOWLEDGMENTS

The research was supported by Swedish Natural Science Council research grant 2320-48. The computing costs were defrayed by grant 104104 of Uppsala Datacentral.

## REFERENCES

Gower, J.C., 1971, A general coefficient of similarity and some of its properties: Biometrics, v. 27, no. 3, p. 857-870.

Hill, M.O., 1974, Correspondence analysis: a neglected multivariate method: Applied statistics, v. 23, no. 3, p. 340-354.

Joreskog, K.G., Klovan, J.E., and Reyment, R.A., 1975, Geological factor analysis: Elsevier Publ. Co., Amsterdam, in press.

Rey, J., Grambast, L., and Ramalho, M., 1974, Données stratigraphiques sur le Crétacé inférieure des environs de Lagos (Algarve, Portugal): Compte Rendu somm. Soc. Géol. Fr., suppl. Tome 16, ser. 3, p. 100-101.

# CLASSIFICATION OF MINNESOTA LAKES BY Q-AND R-MODE FACTOR ANALYSIS OF SEDIMENT MINERALOGY AND GEOCHEMISTRY

Walter E. Dean and Eville Gorham[1]

*Syracuse University and University of Minnesota*

## ABSTRACT

Geographic groupings of Minnesota lakes, defined on the basis of surface-water chemistry, also are distinguishable on the basis of chemical and mineral components of profundal sediments. Q-mode factor analysis of 43 chemical and mineralogical variables in profundal sediments from 46 lakes throughout the State of Minnesota, permitted subdivisions of the water-chemistry groups based mainly on organic- and carbonate-related variables. R-mode factor analysis of the same 43 variables has highlighted many variable associations which were more or less obvious in the raw data. However, many associations not immediately obvious from the raw data were revealed by the R-mode analysis. KEY WORDS: *classification, factor analysis, geochemistry, mineralogy, sedimentology, lakes.*

## INTRODUCTION

Profundal sediments of Minnesota lakes can be characterized on the basis of major chemical and mineral components as well as on the basis of water chemistry (Dean and Gorham, in press). $CaCO_3$ is the main sedimentary variable, and is controlled in part by the marked climatic gradient from northeast to southwest across the State, and in part by the trophic level of the lake. Group 1, or the northeastern Minnesota lakes occur in basins in Precambrian crystalline rocks, have active outlets, and have surface waters of low salinity. Because the profundal sediments of Group 1 lakes

---

[1]Contribution No. 153, Limnological Research Center, University of Minnesota. This work was supported by the National Science Foundation (grants GB 6018 and GB 18800) and the Graduate School, University of Minnesota.

contain no calcium carbonate, they are subdivided mainly on the basis of organic content. The central or Group 2 lakes generally occupy shallow to moderately deep basins in glacial till of differing character, have inactive outlets as a result of net evaporation, and commonly precipitate $CaCO_3$ during the summer months. These lakes can be subdivided on the basis of amounts of organic matter and $CaCO_3$ in the sediments. High-organic Group 2 sediments contain little or no $CaCO_3$ and more than 30 percent organic matter. Sediments of the western Group 2 lakes contain an average of about 38 percent $CaCO_3$, 20 percent organic, and 42 percent clastic. Group 3 lakes occupy shallow depressions in thin glacial drift in the southwestern prairie region of Minnesota. Sediments of these lakes contain the lowest amount of organic matter and only a moderate amount of $CaCO_3$.

The purpose of this paper is to examine the results of analyses of minor chemical components in profundal sediments from Minnesota lakes in regard to lake classification and association of variables obtained from analyses of major chemical and mineral components and surface water chemistry (Dean and Gorham, in press).

## METHODS

Analyses of 43 chemical and mineralogical variables were made on profundal sediments from 46 lakes throughout the State of Minnesota (Fig. 1). Undisturbed samples were collected with a Jenkin corer from the deepest part of each lake, assuming that this sample would provide the best integration of sediment derived from the entire drainage basin and sediment produced within the lake. Only the top 10 cm of sediment from each core was used for analysis, realizing that this does not necessarily represent the same time interval in all cores.

Means and standard deviations of the 43 variables for different lake groups are given in Table 1. Organic matter, carbonate, and clastic components were determined by losses on ignition at 550 and 1,000°C (Dean, 1974). Total concentrations of Ca, Mg, Fe, Mn, Zn, and Cu, and Li were determined by atomic absorption spectrophotometric analyses of solutions obtained by successive digestions of dry lake sediment in HCl, $HNO_3$, and HF. Analyses for total Na and K were made by flame photometry on the same solutions prepared for atomic absorption.

Percent soluble salts is defined as the amount of material removed by leaching the dry sediment sample with 1.0 N ammonium acetate at pH 5.5. This reagent will dissolve calcite and aragonite but not dolomite (Wangersky and Joensuu, 1967). Concentrations of soluble Ca, Mg, Mn, and Fe were determined by atomic absorption on the ammonium acetate leachate. Soluble Na and K were determined by flame photometry on the same ammonium acetate solutions. Sulfur was determined by oxidizing all sulfur compounds with $MnO_2$, extracting sulfate in NaOH, and measuring the sulfate by the ion exchange-conductivity method of Mackereth (1963). Phosphate was determined by the phosphomolydate method after digestion of the sediment with potassium persulfate. Carbon,

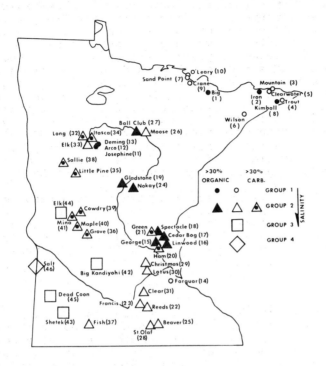

Figure 1. Map of lake location and profundal sediment composition for 46 Minnesota lakes. Lakes are numbered in order of increasing surface-water salinity. Symbols reflect lake groups (1-4) based on surface water chemistry, further modified to separate lakes with profundal sediments rich in calcium carbonate and those rich in organic matter.

hydrogen, and nitrogen were determined chromagraphically with a Hewlett-Packard CHN analyzer. Carbonate carbon was calculated from the ignition loss carbonate determination, and organic carbon calculated as the difference between total carbon and carbonate carbon. Total pigments were measured by techniques described by Gorham and Sanger (1967) and Sanger and Gorham (1970). The sand fraction was separated by wet sieving through a 44µ nylon sieve. The silt and clay fractions were separated by centrifugation. Values in Table 1 for quartz and feldspar in the silt-size fraction, and for calcite, aragonite, dolomite, and quartz in untreated samples are semiquantitative, expressed as x-ray diffraction peak heights in cm. Values for quartz, feldspar, illite, and chlorite in the clay-size fraction also are expressed on a semiquantitative scale of 0 to 4, representing no x-ray diffraction peak to a well-defined peak.

Table 1. Means and standard deviations for 43 chemical and mineral variables in low-organic Group 1 (n=8), high-organic Group 1 and 2 (n=13), intermediate Group 2 (n=10), high-carbonate Group 2 (n=10), and Group 3 (n=5) Minnesota lakes.

| VARIABLE | Low-organic Group 1 | | High-organic Group 1 & 2 | | Intermediate Group 2 | | High-carbonate Group 2 | | Group 3 | |
|---|---|---|---|---|---|---|---|---|---|---|
|  | mean | σ | mean | σ | mean | σ | mean | σ | mean | σ |
| L.O.I. 550 C (%) | 17.96 | 1.74 | 41.92 | 7.09 | 18.19 | 5.63 | 19.84 | 3.90 | 12.40 | 3.78 |
| L.O.I. 1000 C (%) | 3.25 | 0.88 | 5.12 | 3.75 | 17.39 | 6.57 | 38.47 | 5.56 | 20.05 | 2.38 |
| CALCIUM CARBONATE (%) | 78.78 | 2.24 | 52.95 | 6.97 | 63.52 | 5.79 | 41.69 | 5.12 | 67.56 | 5.25 |
| TOTAL Ca (%) | 0.90 | 0.15 | 1.95 | 2.17 | 6.35 | 2.68 | 15.23 | 2.10 | 6.76 | 1.37 |
| TOTAL Mg (%) | 0.73 | 0.18 | 0.43 | 0.10 | 1.02 | 0.26 | 0.85 | 0.28 | 1.67 | 0.66 |
| TOTAL Na (%) | 0.75 | 0.29 | 0.40 | 0.14 | 0.45 | 0.12 | 0.26 | 0.08 | 0.73 | 0.32 |
| TOTAL K (%) | 1.17 | 0.30 | 0.66 | 0.19 | 1.02 | 0.25 | 0.49 | 0.12 | 1.23 | 0.17 |
| TOTAL Fe (%) | 5.01 | 2.14 | 3.80 | 2.17 | 2.86 | 1.54 | 2.21 | 0.76 | 2.46 | 0.52 |
| TOTAL Mn (%) | 0.62 | 1.35 | 0.10 | 0.09 | 0.19 | 0.17 | 0.15 | 0.08 | 0.37 | 0.57 |
| TOTAL Zn (ppm) | 193 | 65 | 155 | 38 | 137 | 48 | 130 | 58 | 130 | 30 |
| TOTAL Cu (ppm) | 75 | 64 | 46 | 55 | 25 | 10 | 38 | 47 | 33 | 11 |
| TOTAL Li (ppm) | 4 | 2 | 2 | 1 | 2 | 1 | 1 | 0 | 3 | 2 |
| AREA (sq. km.) | 4.78 | 4.40 | 2.56 | 4.44 | 1.58 | 1.47 | 2.98 | 2.44 | 7.38 | 8.28 |
| MAX. DEPTH (m) | 29.45 | 21.09 | 11.23 | 6.39 | 15.49 | 8.57 | 17.16 | 9.17 | 4.02 | 2.85 |
| SOLUBLE SALTS (%) | 4.70 | 2.50 | 7.49 | 5.20 | 19.48 | 7.01 | 42.92 | 5.38 | 23.68 | 2.35 |
| SOLUBLE Ca (%) | 0.55 | 0.09 | 1.75 | 1.92 | 5.75 | 3.26 | 15.41 | 2.06 | 6.76 | 1.37 |
| SOLUBLE Mg (%) | 0.08 | 0.03 | 0.12 | 0.06 | 0.50 | 0.14 | 0.53 | 0.20 | 0.63 | 0.23 |
| SOLUBLE Na (ppm) | 2.50 | 0.00 | 5.19 | 2.92 | 5.25 | 2.63 | 7.81 | 1.70 | 8.04 | 5.36 |
| SOLUBLE K (ppm) | 0.29 | 0.11 | 0.34 | 0.16 | 0.52 | 0.17 | 0.41 | 0.09 | 0.86 | 0.55 |
| SOLUBLE Mn (%) | 0.48 | 1.04 | 0.07 | 0.08 | 0.16 | 0.16 | 0.13 | 0.08 | 0.08 | 0.01 |
| SOLUBLE Fe (%) | 0.15 | 0.08 | 0.13 | 0.09 | 0.16 | 0.36 | 0.16 | 0.36 | 0.09 | 0.07 |
| SULFUR (%) | 0.24 | 0.04 | 0.48 | 0.13 | 0.47 | 0.27 | 0.63 | 0.41 | 1.09 | 0.59 |
| PHOSPHORUS (%) | 0.13 | 0.06 | 0.17 | 0.06 | 0.21 | 0.27 | 0.14 | 0.09 | 0.08 | 0.00 |
| NITROGEN (%) | 0.79 | 0.19 | 2.37 | 0.50 | 1.06 | 0.25 | 1.16 | 0.21 | 0.64 | 0.25 |
| TOTAL CARBON (%) | 7.98 | 1.30 | 21.58 | 3.52 | 11.28 | 1.71 | 14.47 | 1.62 | 8.62 | 3.13 |
| HYDROGEN (%) | 1.40 | 0.30 | 3.16 | 0.41 | 1.37 | 0.39 | 1.39 | 0.27 | 1.03 | 0.34 |
| CARBONATE CARBON (%) | 0.39 | 0.11 | 0.60 | 0.47 | 2.09 | 0.80 | 4.63 | 0.67 | 2.50 | 0.40 |
| ORGANIC CARBON (%) | 7.59 | 1.27 | 21.00 | 3.66 | 9.20 | 1.95 | 9.85 | 1.83 | 6.10 | 2.79 |
| TOTAL PIGMENTS (UNITS/GM. ORG. MATTER) | 15.95 | 9.95 | 35.65 | 19.75 | 37.77 | 17.55 | 45.29 | 15.35 | 19.52 | 10.16 |
| SAND (%) | 5.49 | 8.59 | 2.41 | 3.22 | 2.39 | 2.10 | 1.69 | 1.67 | 3.42 | 2.15 |
| SILT (%) | 29.24 | 9.46 | 23.48 | 6.13 | 25.18 | 6.12 | 14.70 | 5.40 | 27.60 | 5.96 |
| CLAY (%) | 30.58 | 8.42 | 20.65 | 6.43 | 25.48 | 8.27 | 13.90 | 2.48 | 23.76 | 5.07 |
| SILT - 4.26Å QUARTZ PK. | 2.86 | 1.65 | 3.01 | 1.94 | 5.56 | 1.41 | 4.01 | 1.65 | 6.20 | 0.96 |
| SILT - 3.20Å F'SPAR PK. | 1.18 | 1.03 | 1.97 | 1.45 | 1.73 | 1.70 | 3.35 | 1.86 | 1.57 | 1.02 |
| SILT - 3.18Å F'SPAR PK. | 2.76 | 1.88 | 3.65 | 2.32 | 3.45 | 2.52 | 4.60 | 1.72 | 2.64 | 0.81 |
| CLAY - ILLITE PEAK | 1.25 | 0.89 | 0.54 | 0.52 | 2.00 | 0.94 | 1.10 | 0.88 | 2.40 | 0.55 |
| CLAY - CHLORITE PEAK | 1.63 | 0.52 | 0.62 | 0.87 | 1.80 | 0.63 | 1.10 | 0.88 | 2.60 | 0.55 |
| CLAY - QUARTZ PEAK | 1.63 | 0.52 | 1.38 | 0.77 | 1.90 | 0.32 | 1.60 | 0.70 | 2.00 | 0.00 |
| CLAY - FELDSPAR PEAK | 0.88 | 0.64 | 0.85 | 0.90 | 1.30 | 0.82 | 1.10 | 0.88 | 1.40 | 0.55 |
| X-RAY CALCITE PEAK (cm) | 0.00 | 0.00 | 1.56 | 3.56 | 7.49 | 5.43 | 22.61 | 3.14 | 14.68 | 7.78 |
| X-RAY ARAGONITE PEAK (cm) | 0.00 | 0.00 | 0.00 | 0.00 | 0.10 | 0.32 | 0.38 | 0.54 | 0.51 | 0.72 |
| X-RAY DOLOMITE PEAK (cm) | 0.00 | 0.00 | 0.21 | 0.77 | 1.70 | 1.00 | 0.54 | 0.49 | 3.10 | 1.77 |
| X-RAY QUARTZ PEAK (cm) | 13.27 | 7.93 | 9.38 | 6.59 | 12.19 | 5.69 | 6.40 | 2.67 | 16.76 | 3.52 |

Grouping of the 46 lakes based on the 43 sediment variables listed in Table 1 was done using a FORTRAN IV Q-mode factor analysis program developed at the University of Kansas by Ondrick and Srivastava (1970). For the Q-mode analysis, all values were standardized by dividing the value for each variable by the maximum value of each variable occurring in the 46 lakes studied. The resulting input matrix consisted of 43 converted variables, each with a range of 0 to 1 and 46 observations (lake sediment samples). A total of 10 factors was extracted, although 84 percent of the variance in the data was explained by the first 3 factors, after varimax rotation (Table 2).

Relationships among variables were examined by R-mode factor analysis using the same program as for the Q-mode analysis (Ondrick and Srivastava, 1970). Loadings for 10 factors extracted, after varimax rotation, are given in Table 3.

Table 2. Loadings for first three rotated factors of Q-mode factor analysis, 43 variables, 46 lakes.

| LAKE NO. | LAKE NAME | COUNTY | F 1 | F 2 | F 3 |
|---|---|---|---|---|---|
| 1 | BIG | ST. LOUIS | 0.60345 | -0.25198 | -0.61156 |
| 2 | IRON | COOK | 0.26247 | -0.22820 | -0.82484 |
| 3 | MOUNTAIN | COOK | 0.58759 | -0.20210 | -0.41974 |
| 4 | TROUT | COOK | 0.52901 | -0.20785 | -0.56661 |
| 5 | CLEARWATER | COOK | 0.65860 | -0.18552 | -0.36700 |
| 6 | WILSON | LAKE | 0.63501 | -0.18903 | -0.43717 |
| 7 | SAND POINT | ST. LOUIS | 0.42912 | -0.21576 | -0.40365 |
| 8 | KIMBALL | COOK | 0.39859 | -0.31523 | -0.82926 |
| 9 | CRANE | ST. LOUIS | 0.71508 | -0.27106 | -0.40715 |
| 10 | O'LEARY | ST. LOUIS | 0.82733 | -0.24639 | -0.40628 |
| 11 | JOSEPHINE | HUBBARD | 0.65032 | -0.31187 | -0.63607 |
| 12 | ARCO | HUBBARD | 0.65224 | -0.31004 | -0.65512 |
| 13 | DEMING | HUBBARD | 0.59554 | -0.31146 | -0.70242 |
| 14 | FARQUAR | DAKOTA | 0.81691 | -0.22104 | -0.38636 |
| 15 | GEORGE | ANOKA | 0.50240 | -0.26383 | -0.75567 |
| 16 | LINWOOD | ANOKA | 0.26216 | -0.52304 | -0.72337 |
| 17 | CEDAR BOG | ANOKA | 0.14875 | -0.30208 | -0.90918 |
| 18 | SPECTACLE | ISANTI | 0.38767 | -0.29984 | -0.82440 |
| 19 | GLADSTONE | CROW WING | 0.28720 | -0.24551 | -0.79342 |
| 20 | HAM | ANOKA | 0.08439 | -0.86595 | -0.39194 |
| 21 | GREEN | ISANTI | 0.22434 | -0.80937 | -0.39160 |
| 22 | REEDS | WASECA | 0.72917 | -0.38332 | -0.45479 |
| 23 | FRANCIS | LESUEUR | 0.73155 | -0.44729 | -0.43137 |
| 24 | NOKAY | CROW WING | 0.31676 | -0.55057 | -0.69510 |
| 25 | BEAVER | STEELE | 0.63380 | -0.64233 | -0.27292 |
| 26 | MOOSE | ITASCA | 0.56858 | -0.72369 | -0.22782 |
| 27 | BALL CLUB | ITASCA | 0.29522 | -0.35554 | -0.66278 |
| 28 | ST. OLAF | WASECA | 0.72782 | -0.54710 | -0.29204 |
| 29 | CHRISTMAS | HENNEPIN | 0.76688 | -0.40694 | -0.36576 |
| 30 | LOTUS | CARVER | 0.72936 | -0.54175 | -0.30433 |
| 31 | CLEAR | LESUEUR | 0.77783 | -0.47690 | -0.35160 |
| 32 | LONG | CLEARWATER | 0.45773 | -0.82645 | -0.13542 |
| 33 | ELK | CLEARWATER | 0.13841 | -0.58876 | -0.53734 |
| 34 | ITASCA | CLEARWATER | 0.09715 | -0.87396 | -0.38396 |
| 35 | LITTLE PINE | OTTERTAIL | 0.25492 | -0.90033 | -0.21657 |
| 36 | GROVE | POPE | 0.43708 | -0.83224 | -0.27663 |
| 37 | FISH | COTTONWOOD | 0.54875 | -0.59213 | -0.33193 |
| 38 | SALLIE | BECKER | 0.39223 | -0.86605 | -0.20811 |
| 39 | COWDRY | DOUGLAS | 0.34805 | -0.78588 | -0.35327 |
| 40 | MAPLE | DOUGLAS | 0.42866 | -0.85186 | -0.19893 |
| 41 | MINA | DOUGLAS | 0.30015 | -0.87741 | -0.26467 |
| 42 | SHETEK | MURRAY | 0.55552 | -0.65275 | -0.24636 |
| 43 | BIG KANDIYOHI | KANDIYOHI | 0.61062 | -0.61183 | -0.22991 |
| 44 | ELK | GRANT | 0.60440 | -0.60897 | -0.20041 |
| 45 | DEAD COON | LINCOLN | 0.82015 | -0.47516 | -0.14819 |
| 46 | SALT | LACQUI PARLE | 0.76219 | -0.44318 | -0.21532 |
| | | | 13.81637 | 13.68129 | 11.13719 |
| | | | 30.03557 | 29.74191 | 24.21127 |

Table 3. Loadings for 10 rotated factors of R-mode factor analysis, 43 variables, 46 lakes.

THE ROTATED MATRIX OF FACTOR LOADINGS

| VARIABLES | F 1 | F 2 | F 3 | F 4 | F 5 | F 6 | F 7 | F 8 | F 9 | F 10 |
|---|---|---|---|---|---|---|---|---|---|---|
| 1 L.O.I. 550° C | -0.26668 | 0.90177 | 0.08142 | -0.03841 | -0.02326 | -0.02917 | 0.05053 | -0.04884 | -0.27099 | 0.08662 |
| 2 CARBONATE | -0.94529 | -0.22127 | -0.05111 | -0.15465 | -0.05934 | -0.02440 | -0.06659 | -0.05508 | -0.02592 | -0.08111 |
| 3 CLASTIC | -0.73958 | 0.56869 | -0.12607 | -0.21961 | -0.02014 | -0.00770 | -0.10131 | 0.00508 | -0.05840 | -0.07311 |
| 4 TOTAL Ca | -0.94777 | 0.20555 | -0.07319 | -0.15430 | -0.06133 | -0.04755 | -0.05649 | -0.00114 | -0.05840 | -0.00858 |
| 5 TOTAL Mg | -0.20513 | 0.40478 | -0.02856 | -0.00462 | -0.20554 | -0.00348 | -0.10348 | -0.05123 | 0.77206 | -0.08501 |
| 6 TOTAL Na | -0.57388 | -0.47575 | -0.04592 | -0.04470 | -0.21002 | -0.05907 | -0.05705 | -0.24774 | -0.12846 | 0.05284 |
| 7 TOTAL K | -0.61416 | -0.55923 | -0.04264 | 0.06075 | -0.16505 | -0.11176 | -0.01575 | -0.02347 | -0.44420 | -0.01404 |
| 8 TOTAL Fe | -0.29648 | -0.03086 | -0.42771 | 0.21614 | -0.13155 | -0.02929 | -0.03647 | -0.05834 | -0.44670 | -0.03051 |
| 9 TOTAL Mn | -0.03637 | -0.10526 | -0.41842 | -0.01373 | 0.05764 | -0.06213 | -0.02998 | -0.13108 | -0.06877 | -0.10476 |
| 10 TOTAL Zn | -0.36644 | -0.08273 | -0.46758 | -0.21658 | -0.04203 | -0.28731 | -0.66762 | -0.45571 | -0.01803 | -0.11603 |
| 11 TOTAL Cu | -0.12603 | -0.01512 | -0.11160 | -0.10756 | -0.25750 | -0.05441 | -0.01225 | -0.12239 | -0.08451 | 0.04707 |
| 12 TOTAL Li | -0.61089 | -0.23763 | -0.05508 | -0.20154 | -0.29701 | -0.13596 | -0.08500 | -0.09775 | -0.02983 | -0.06407 |
| 13 AREA | -0.02874 | -0.17023 | -0.13929 | -0.01227 | -0.03841 | -0.01939 | -0.02983 | -0.31371 | -0.16910 | 0.08202 |
| 14 MAXIMUM DEPTH | -0.01835 | -0.23406 | -0.60066 | -0.15726 | -0.38411 | -0.00734 | -0.06308 | -0.01676 | -0.06713 | -0.19258 |
| 15 SOLUBLE SALTS | -0.95644 | -0.20555 | -0.01140 | -0.12739 | -0.05485 | -0.01258 | -0.06388 | -0.01686 | -0.00890 | -0.01525 |
| 16 SOLUBLE Ca | -0.95214 | -0.29074 | -0.05604 | -0.14039 | -0.05485 | -0.05577 | -0.01260 | -0.03211 | -0.52930 | -0.00789 |
| 17 SOLUBLE Mg | -0.60978 | -0.29197 | -0.01140 | -0.14935 | -0.24376 | -0.04308 | -0.11211 | -0.31568 | -0.12986 | -0.11489 |
| 18 SOLUBLE Na | -0.60289 | -0.16949 | -0.10269 | -0.12600 | -0.09911 | -0.44308 | -0.05525 | -0.10077 | -0.35549 | -0.25733 |
| 19 SOLUBLE K | -0.05419 | -0.10054 | -0.01028 | -0.09789 | -0.07867 | -0.04999 | -0.09117 | -0.14287 | -0.14399 | -0.12759 |
| 20 SOLUBLE Mn | -0.01755 | -0.09834 | 0.97401 | 0.00603 | -0.03378 | -0.08175 | -0.00342 | -0.02297 | -0.52853 | -0.01906 |
| 21 SOLUBLE Fe | -0.02171 | -0.04126 | -0.22219 | -0.17797 | -0.06024 | -0.03758 | -0.23488 | -0.02449 | -0.11849 | -0.20121 |
| 22 TOTAL SULFUR | -0.21343 | -0.01565 | -0.03981 | -0.16481 | -0.80344 | -0.10734 | -0.06838 | -0.05881 | -0.17228 | -0.02951 |
| 23 TOTAL PHOSPHORUS | -0.16262 | -0.03685 | -0.07905 | -0.10898 | -0.00046 | -0.10798 | -0.08449 | -0.01768 | -0.06275 | -0.07603 |
| 24 TOTAL NITROGEN | -0.05448 | 0.93074 | -0.13589 | -0.08291 | -0.01275 | -0.11739 | -0.03540 | -0.05144 | -0.16407 | -0.03385 |
| 25 TOTAL CARBON | -0.05410 | 0.94978 | -0.01402 | -0.11135 | -0.06520 | -0.00063 | -0.01597 | -0.02482 | -0.24204 | -0.03687 |
| 26 TOTAL HYDROGEN | -0.28216 | 0.89773 | 0.05658 | -0.00447 | -0.01816 | -0.02507 | -0.04212 | -0.02021 | -0.74174 | -0.03376 |
| 27 CARBONATE CARBON | -0.94215 | -0.22884 | -0.02219 | -0.15782 | -0.06024 | -0.06896 | -0.00684 | -0.03477 | -0.52748 | -0.00609 |
| 28 ORGANIC CARBON | -0.30535 | 0.93827 | -0.11082 | -0.06102 | -0.17817 | -0.11134 | -0.29061 | -0.18725 | -0.44866 | -0.10476 |
| 29 TOTAL PIGMENTS | -0.12159 | -0.23408 | -0.10785 | -0.07524 | -0.07511 | -0.12335 | -0.07870 | -0.17585 | -0.00764 | -0.10602 |
| 30 % SAND | -0.49417 | -0.15435 | -0.33750 | -0.21000 | -0.14814 | -0.08634 | -0.00705 | -0.65874 | -0.26725 | -0.10607 |
| 31 % SILT | -0.12159 | -0.13316 | 0.45761 | -0.07713 | -0.10631 | -0.04061 | -0.03500 | -0.01664 | -0.11453 | -0.13015 |
| 32 % CLAY | -0.61708 | -0.41226 | 0.19346 | -0.47713 | -0.07438 | -0.08212 | -0.26163 | -0.08158 | -0.19188 | -0.19188 |
| 33 4.26 A QUARTZ (SILT) | -0.00603 | -0.27629 | -0.17915 | -0.00752 | -0.07482 | -0.18212 | -0.16674 | -0.08036 | -0.55200 | -0.19188 |
| 34 4.26 A F'SPAR (SILT) | 0.34648 | -0.14981 | -0.17961 | -0.49703 | -0.24047 | -0.03547 | -0.07009 | -0.03799 | -0.01521 | -0.09451 |
| 35 4.18 A F'SPAR (SILT) | -0.17170 | -0.14102 | -0.03486 | 0.87653 | -0.05477 | -0.00877 | -0.05197 | -0.03290 | -0.05973 | -0.04750 |
| 36 ILLITE (CLAY) | -0.13011 | -0.45752 | -0.03977 | -0.88054 | -0.01877 | -0.20156 | -0.05197 | -0.01597 | -0.74174 | -0.10524 |
| 37 CHLORITE (CLAY) | -0.07395 | 0.49478 | -0.00817 | -0.05085 | -0.00202 | -0.16934 | -0.04142 | -0.20218 | -0.52748 | -0.00603 |
| 38 QUARTZ (CLAY) | -0.20221 | 0.38947 | -0.12585 | -0.05809 | -0.00904 | -0.18975 | -0.29066 | -0.31475 | -0.44866 | -0.32405 |
| 39 FELDSPAR (CLAY) | -0.05338 | -0.17667 | -0.12626 | -0.20616 | -0.20131 | -0.07473 | -0.20131 | -0.70968 | -0.01803 | -0.07416 |
| 40 CALCITE (X-RAY) | -0.90152 | -0.15662 | -0.07908 | -0.27297 | -0.00085 | -0.07473 | -0.01204 | -0.16048 | -0.22797 | -0.05855 |
| 41 ARAGONITE (X-RAY) | -0.34470 | -0.02071 | -0.05631 | -0.06647 | -0.29463 | -0.03448 | -0.12684 | -0.06898 | -0.44099 | -0.05715 |
| 42 DOLOMITE (X-RAY) | -0.48145 | -0.22078 | -0.05242 | -0.18204 | -0.24063 | -0.03248 | -0.00968 | -0.10840 | -0.76609 | -0.09715 |
| 43 QUARTZ (X-RAY) | -0.48145 | -0.29527 | -0.24885 | -0.02322 | -0.13913 | -0.13957 | -0.18587 | -0.05444 | -0.41833 | -0.17933 |
| SUM SQ | 9.51063 | 6.99327 | 3.40811 | 2.64046 | 2.00678 | 2.04103 | 1.73887 | 1.56117 | 5.02992 | 1.40057 |
| VAREXP | 22.11774 | 16.26341 | 7.92584 | 6.14061 | 4.66693 | 4.74658 | 4.04388 | 3.63063 | 11.69749 | 3.25714 |

## LAKE CLASSIFICATION - Q-MODE FACTOR ANALYSIS

Loadings for the first three rotated factors from the Q-mode analysis, listed in Table 2, are plotted as a triaxial diagram in Figure 2. The lake groupings resulting from this analysis are essentially the same as the groupings obtained by Q-mode analysis of 27 major chemical and mineralogical variables on the same 46 lakes (Dean and Gorham, in press). The groupings are delineated mainly on the basis of organic-, carbonate-, and clastic-related variables. In addition to representing loadings for the three main Q-mode factors, the triaxial diagram in Figure 2 has been modified further to represent lake groups based on surface-water salinity (size of circles), and lakes grouped on the basis of high-organic sedimentary matter (black circles) and high sedimentary carbonate (stippled circles). Figure 2 shows

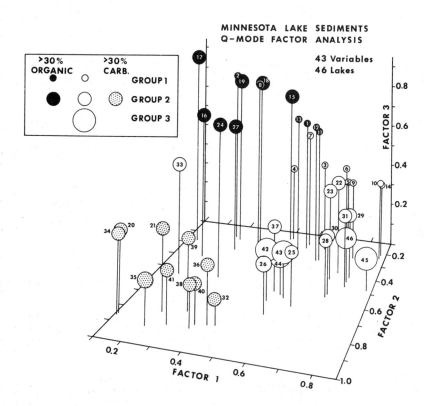

Figure 2. Triaxial plot of factor loadings for first 3 factors of Q-mode factor analysis (Table 2). Size of circles is indicative of surface-water salinity (water chemistry groups 1, 2, and 3). Black circles represent lakes with greater than 30 percent organic matter in profundal sediments; stippled circles represent lakes with greater than 30 percent calcium carbonate in profundal sediments.

that factor 3 clearly delineates lakes based on organic-related variables (e.g. ignition loss organic matter, organic carbon, hydrogen, nitrogen). Factor 2 delineates lakes mainly on the basis of sedimentary carbonate variables (e.g. ignition loss carbonate, soluble salts, soluble Ca, Mg, Na, and K, and x-ray calcite, aragonite, and dolomite). Factor 1 delineates lakes mainly on the basis of the negative relationships between clastic variables (e.g. ignition loss clastic, Na, K, Li, sand, silt, clay, and clay mineralogy) and both carbonate- and organic-related variables. Factor 1 therefore tends to separate further groups based on factors 2 and 3 into the third dimension in Figure 2. These negative relationships are the result of two dilution trends beginning with an inorganic, noncarbonate clastic base. In central and northeastern Minnesota, dilution of the clastic base is mainly by addition of autochthonous and allochthonous organic matter (Gorham and Sanger, 1975; Sanger and Gorham, 1970). Most of these lakes do not become saturated with respect to $CaCO_3$ even during the summer months. Factor 3 therefore subdivides Group 1 and Group 2 lakes (based on surface water salinity) into those with greater than 30 percent organic matter in their profundal sediments, and those with less than 30 percent organic matter.

In west-central Minnesota, an excess of evaporation over precipitation increases the salinity of lake surface waters to the extent that commonly they precipitate calcium carbonate during the summer months. Profundal sediments of these western Group 2 lakes usually contain more than 30 percent calcium carbonate. In Figure 2, the high carbonate, western Group 2 lakes clearly stand out as a result of high loadings on factor 2 and low loadings on factor 1. Elk Lake (lake number 33) is an intermediate between high organic and high carbonate Group 2 lakes and illustrates the main problem of classifying lakes on the basis of precipitated calcium carbonate. Elk Lake is one of the deepest Group 2 lakes studied (30 m), and the sediments in the deepest portion of the lake are varved, suggesting anoxic conditions throughout most of the year. Surface sediments at 30 m contain only 19 percent calcium carbonate, although sediments from 5 other stations, ranging in depth from 11-24 m, contain 24 to 70 percent calcium carbonate. Most sediments in Elk Lake would classify it as a high-carbonate western Group 2 lake, but dissolution of carbonate in the deeper waters has greatly reduced the amount of carbonate being incorporated into the sediments there.

Lakes of the western and southwestern prairie regions of Minnesota can be distinguished as a distinct third group (Group 3 in Figs. 1 and 2) on the basis of their higher surface water salinities. However, on the basis of profundal sediment characteristics, they classify with the intermediate Group 2 lakes (Fig. 2). The Group 3 lakes occupy shallow depressions in thin, gypsum-bearing glacial drift derived partly from Cretaceous shales. Most of these lakes are highly productive, but organic matter in the profundal sediments is generally low (<16%) as a combined result of oxidation of organic matter in shallow, wind-mixed waters, and dilution by high rate of silting. Dilution by high rate of clastic influx is probably also responsible for the relatively low calcium-carbonate content of Group 3 sediments, even though these lakes are more saline than the western Group 2 lakes. Consequently, Group 3 lakes are characterized by a predominance of clastic material, only a moderate percentage of carbonate, and a small percentage of organic matter, which is diluted by both clastic and carbonate deposition.

On the basis of water chemistry, a rare fourth group of lakes, characterized by saline (conductivity greater than 6,000 µmhos/cm$^2$, 25°C) sodium-sulfate waters, occur along the Minnesota and North and South Dakota borders (Fig. 1). Group 4 lakes are more common in the Dakotas, and are represented in this study by a single example, Salt Lake (lake no. 46; Figs. 1 and 2). In terms of sediment characteristics, Salt Lake is indistinguishable from Group 3 lakes and has been included with them in the sediment classification.

## ASSOCIATIONS OF VARIABLES - R-MODE FACTOR ANALYSIS

Loadings for 10 factors, after varimax rotation, extracted by the R-mode factor analysis are given in Table 3. The factor loadings are actually summaries of associations based on the correlation coefficient matrix of 43 variables and 46 observations (lakes) which was used as input to the factor-analysis program of Ondrick and Srivastava (1970). The R-mode analysis was checked using a factor analysis program (BMD-X72) in the University of California Biomedical Data statistical package (Dixon, 1970). Both analyses gave almost identical results.

Factor 1, accounting for about 22 percent of the variance in the data, is based on the positive associations among "carbonate" variables (e.g. ignition loss CaCO$_3$, total Ca, soluble salts, soluble Ca, Na, and Mg, and x-ray calcite-peak height), and the negative relationships between "carbonate" variables and certain "clastic" variables (e.g. ignition loss clastic, total Na, K, and Li, % silt, % clay, and x-ray quartz peak height).

Factor 2, accounting for 16.3 percent of the variance, is based on negative correlations between "organic" variables (e.g. ignition loss organic matter, carbon, hydrogen, and nitrogen) and certain "clastic" variables (e.g. ignition loss clastic, total Na and K, % clay, illite, and chlorite). Notice that clastic variables contribute to variance in two factors based on negative correlations with carbonate (factor 1) and organic (factor 2). This is a reflection of the organic and carbonate dilution trends mentioned earlier, that is beginning with an inorganic, noncarbonate clastic base, Minnesota lakes seem to show two distinct trends, one in carbonate enrichment and one in organic enrichment, with few lakes showing enrichment in both carbonate and organic matter. Table 3 shows that the inverse association between carbonate and clastic is somewhat closer than the inverse association between organic and clastic. The clay mineral variables illite, chlorite, and clay-size quartz are correlated negatively with organic matter but not with carbonate. However, total clastic material, as represented by % clastic, % silt, % clay, and total Na and K, shows negative correlations with both organic and carbonate variables.

Factor 3, accounting for 8 percent of the variance, is essentially an iron-manganese factor, and relates those variables most affected by redox conditions in the profundal region. Highest loadings are for soluble and total Mn, followed by maximum depth. A significant result of the R-mode analysis is the association of depth with only the redox sensitive elements iron, manganese, and zinc. Because these elements are more mobile under reducing conditions, the positive association between these elements and depth probably implies that greater depth is associated with increased reducing conditions and therefore increased concentrations of the redox-sensitive elements. Elk Lake (lake number 33, Fig. 1) is particularly instructive in this regard. The same increase in reducing conditions in the deeper portions of Elk Lake which produced dissolution of carbonate (discussed earlier) also has produced a marked increase in iron, manganese, and sulfur in the deepest profundal sediments (Dean, unpublished data). Manganese, which is most sensitive to redox variations (e.g. Krauskopf, 1957; Hem, 1972), would be most closely associated with depth, and therefore has the highest loadings on factor 3.

Factor 9, accounting for 11.7 percent of the variance after varimax rotation, is the result of associations of magnesium with both dolomite and clay minerals. In factor 1, soluble Mg is associated with the carbonate variables. Because the ammonium acetate extraction technique does not dissolve dolomite (Wangersky and Joensuu, 1967), this Mg contribution must represent Mg coprecipitated with $CaCO_3$. However, in factor 9, highest loadings are for total Mg, x-ray dolomite, illite, and chlorite. In the western (high carbonate) profundal sediments of Group 2, much of the Mg is in the form of dolomite. The third source of Mg, that is clay minerals, especially chlorite, is most significant in the noncarbonate sediments of Group 1 lakes in northeastern Minnesota where glacial drift derived from Precambrian metamorphic rocks would be expected to contribute relatively large amounts of Mg-rich minerals to the clastic fraction.

The four rotated factors discussed above account for 58 percent of the variance in the data, and incorporate most of the relationships among variables. Each of the remaining 6 factors contributes less than 7 percent of the variance. All 10 factors account for 84.5 percent of the total variance, leaving 14.5 percent unexplained by the 10 factors. Factor 4 is mainly the result of correlations among the silt variables (% silt, silt-size quartz and feldspar). Factor 5 is the result of a correlation between sulfur and soluble K. Factor 6 is based on a weak negative correlation (-0.42) between phosphorus and clay-size feldspar. Factor 7 suggests a negative relationship between total pigments and total copper. Factor 8 suggests negative relationships between Zn and Cu and % sand. Factor 10 is based mainly on area.

## REFERENCES

Dean, W.E., 1974, Determination of carbonate and organic matter in calcareous sediments and sedimentary rocks by loss on ignition: comparison with other methods: Jour. Sed. Pet., v. 44, no. 1, p. 242-248.

Dean, W.E., and Gorham, E., Major chemical and mineral components of profundal surface sediments in Minnesota lakes: Limnol. Oceanog. (in press).

Dixon, W.J., ed., 1970, BMD Biomedical computer programs, X-series supplement (2nd ed.): Univ. Calif. Publ. in Automatic Computation, no. 2, 260 p.

Gorham, E., and Sanger, J.E., 1967, Plant pigments in woodland soils: Ecology, v. 48, p. 306-308.

Gorham, E., and Sanger, J.E., 1975, Fossil pigments in Minnesota lake sediments, and their bearing upon the balance between terrestrial and aquatic inputs to sedimentary organic matter: Intern. Ver. Theoret. Angew, Limnol. Verhandl. (in press).

Hem, J.D., 1972, Chemical factors that influence the availability of iron and manganese in aqueous systems: Geol. Soc. America Bull., v. 83, no. 2, p. 443-450.

Krauskopf, K.B., 1957, Separation of manganese from iron in sedimentary processes: Geochim. et Cosmochim. Acta, v. 12, nos. 1/2, p. 61-84.

Mackereth, F.J.H., 1963, Some methods of water analysis for limnologists: Sci. Publ. Freshwater Biol. Assoc., no. 21, 71 p.

Ondrick, C.W., and Srivastava, G.S., 1970, CORFAN-FORTRAN IV computer program for correlation, factor analysis (R- and Q-mode) and varimax rotation: Kansas Geol. Survey Computer Contr. 42, 92 p.

Sanger, J.E., and Gorham, E., 1970, The diversity of pigments in lake sediments and its ecological significance: Limnol. Oceanog., v. 15, p. 59-69.

Wangersky, P.J., and Joensuu, O.I., 1967, The fractionation of carbonate in deep-sea cores: Jour. Geology, v. 75, no. 2, p. 148-177.

# SEDIMENTARY ENVIRONMENTAL ANALYSIS OF LONG ISLAND SOUND, USA WITH MULTIVARIATE STATISTICS

Peter H. Feldhausen and Syed A. Ali

*Dames & Moore and SUNY Stony Brook*

ABSTRACT

A multivariate statistical strategy, which reduces the sample-variable matrix to a set of interpretable graphical relationships, was used to maximize the environmental information extracted from bottom-sediment, grain-size data for Long Island Sound, USA. The weight-percent whole phi variables were tested for redundancy using R-mode cluster analysis. Q-mode cluster analysis partitioned 57 traverse samples into five facies. This classification was extended to the other 171 samples through discriminant analysis. Ordination was employed to depict the gradational relationships among the samples and facies, and to observe significant environmental and textural parameter gradients within the sample space. Interpretations obtained with the ordination and with other standard techniques were tested by comparing the facies map with the distribution of known environmental phenomena.

Five environmentally significant facies were determined for Long Island Sound with the described strategy. These facies are: (1) clayey silt, (2) sandy-clayey silt, (3) silty-clayey sand, (4) silty sand, and (5) sands, which contain individually somewhat dissimilar samples. The sand and silty-sand facies are restricted to shoal areas and the margins of the Sound. Most of Long Island Sound, however, is blanketed by sediments which contain a high portion of silt-sized particles. A scarcity of sediment sources is suggested by the fine nature of the sediment.
KEY WORDS: *cluster analysis, discriminant analysis, ordination, principal components analysis, trend analyses, marine sediments, oceanography.*

## INTRODUCTION

A promising development in the field of sedimentology has been the application of robust multivariate statistical techniques which afford an objective means of classifying and comparing sedimentary environments. Imbrie and Purdy (1962) employed factor analysis in their study of modern Bahamian carbonates. Factor analysis also has been applied to advantage in determining the depositional environments of sediments from their grain-size distribution (Klovan, 1966; Swift and others, 1971; Knebel and Creager, 1973).

This study, however, chooses to employ a multivariate strategy previously used in paraecologic and paleoenvironmental studies (Park and Feldhausen, 1969; Park, 1974), and recently used to characterize modern sedimentary environments (Feldhausen and Ali, 1974; Ali, Lindemann, and Feldhausen, 1975). Cluster analysis, ordination, and discriminant analysis are applied sequentially to extract sedimentological information from phi weight percent data for 57 bottom samples obtained by M.A. Buzas (1974, written communication) from Long Island Sound, USA (Fig. 1). The main points of this strategy, portrayed graphically in Figure 2, are to: (1) test whole phi variables for redundancy using R-mode cluster analysis; (2) classify the 57 samples into environmentally significant classes or facies using Q-mode cluster analysis; (3) interpret the classification using a three-dimensional, Q-mode ordination and gradient analysis on a two-dimensional, Q-mode ordination; (4) test the significance of the classification using Wilks' lambda analysis; (5) verify the classification and ordination by comparing the distribution of the samples within each model with their field relations; and (6) extend the classification with the aid of discriminant analysis to 171 additional bottom samples obtained from Long Island Sound.

## THE STUDY AREA

A complicated bottom topography lies beneath the waters of Long Island Sound. The complex of small basins and intervening sills (Fig. 1) affects the circulation of the subsurface waters and, hence, the movements of bottom sediment.

The western section of the Sound is known as the Narrows. Hell's Gate, a rock sill at a depth of 25 m forms the westernmost boundary. At the eastern margin of the Narrows is the Hempstead Sill, where the water depth is less than 12 m. Between the Hempstead Sill and Stratford Shoals is the Western Basin, an area which is deeper and wider than the Narrows. The bottom of this basin also is irregular; a north-south shoal area exists in the middle of the basin. It extends from Eaton's Neck through Cable and Anchor Reefs to Sheffield Island. A narrow, deep passage just south of Cable and Anchor Reefs permits bottom waters to flow between both halves of the Western Basin.

The Central Basin is bounded on the west by the Stratford Shoals and on the east by the Mattituck Sill. Stratford Shoals

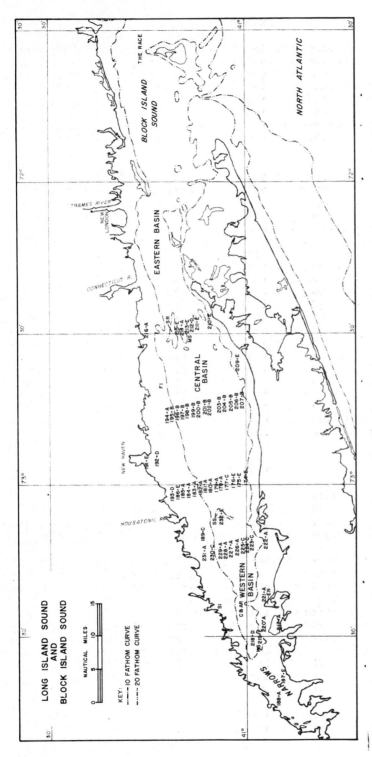

Figure 1. Chart of Long Island Sound showing bottom-sampling stations occupied by Buzas (1965). HS-Hempstead Sill; EN-Eatons Neck Point; C&AR-Cable and Anchor Reefs; SI-Sheffield Island; SS-Stratford Shoals; FI-Fishers Island; MS-Mattituck Sill; SR-Six Mile Reef.

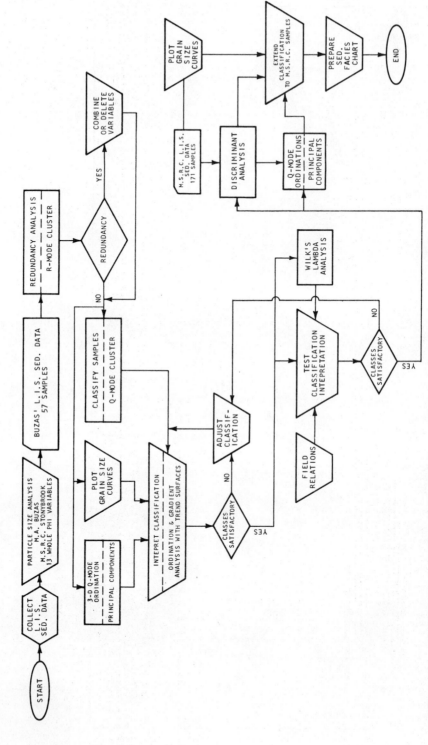

Figure 2. Flowchart of multivariate statistical strategy employed in study of Long Island Sound sediments.

# SEDIMENTARY ENVIRONMENTAL ANALYSIS

do not restrict the flow of bottom water, whereas the Mattituck Sill, with water depths less than 25 m, partially restricts communication between the Central and Eastern Basins.

The southern shore of Long Island Sound along the Eastern Basin is devoid of inlets and bays as it is in the Central Basin. Conversely, the northern shore of both basins is punctuated by the large harbors. Most of the freshwater and river-borne sediment enter Long Island Sound through the Connecticut and Thames Rivers which drain a large portion of glaciated New England.

Long Island Sound is characterized by minimum tidal range and maximum tidal currents at its connection with Block Island Sound and by maximum tidal range and minimum tidal currents at the Narrows. Tidal currents flowing through the Race may reach 5 knots, but they seldom reach 1 knot at Hell's Gate. Tidal-current profiles obtained by Riley (1956, fig. 7) off Six Mile Reef near the boundary between the Central and Eastern Basins indicate that tidal currents may exceed 1.6 knots at depths of 30 m.

Seabed-drifter data (Gross and Bumpus, 1972; Hollman and Sandberg, 1972) suggest that the near-bottom currents are stronger in the eastern than in the western portions of Long Island Sound. The net drift speed ranged between 1.2 cm/sec south of the Thames River estuary to 0.5 cm/sec in the Western Basin. Net drift speed on a north-south transect across the Sound at Six Mile Reef was 0.8 cm/sec. East-west facing symmetrical and asymmetrical megaripples and current bedding, which occur south of the Connecticut River estuary and across the Mattituck Sill (Grim, Drake, and Heirtzler, 1970) are a manifestation of the strong bottom currents in this area.

In the Eastern Basin the residual bottom drift (Gross and Bumpus, 1972) was to the west with a northerly component toward the Connecticut shore. However, in the Central and Western basins the northward component was less pronounced and the drift was primarily westward.

Water circulation in the Central Basin is less dynamic than in the Eastern Basin. The Central Basin is not known to exhibit megaripples (Grim, Drake, and Heirtzler, 1970). Within the basin there is decreasing transport in the bottom-water layers because of upwelling and nearshore mixing. According to Gordon and Pilbeam (1973), the bottom water flows into the Central Basin principally along the north side of Six Mile Reef where it becomes mixed with Connecticut River water. Inshore of the 20-m contour, the flow is toward the coast. Hardy (1972) has characterized the waters of the Central Basin as seasonally homogeneous. However, winter-formed bottom water exhibits a seasonal lag in warming. The persistence into summer of this colder and denser bottom layer indicates limited mixing. Mixing, however, between bottom and surface waters does occur along both shores and in the vicinity of Six Mile Reef (Gordon and Pilbeam, 1973).

Water flows into western Long Island Sound from both directions. Bottom water enters from the east principally in the area south of Stratford Shoal, while relatively freshwater enters on the west from New York Harbor.

Because of limited fetch and generally light winds (Fig. 1 and Table 1), wave action on Long Island Sound is not strong. On the average, the significant wave heights before shoaling are estimated to be less than 0.5 m. This suggests that average wave base is at a depth of 6 m or less. Sustained winds from a strong gale, on the other hand, are estimated to generate significant waves about 3 m in height with a corresponding wave base at depths of 25 to 30 m. Records of severe local storms (Pautz, 1969) suggest that storms of this intensity occur, on the average, less than once per year in the vicinity of Long Island Sound.

Table 1. Mean average wind velocity, Long Island Sound[1].

| Month | Mean Direction | Mean Speed (mph) |
|---|---|---|
| January | N | 8 |
| February | N | 8 |
| March | NW | 9 |
| April | WSW | 12 |
| May | S | 7 |
| June | S | 8 |
| July | S | 6 |
| August | S | 7 |
| September | WSW | 10 |
| October | N | 8 |
| November | N | 8 |
| December | NNW | 10 |

[1]Values obtained by interpolation of data presented in Climatic Atlas of the United States: Environmental Science Services Administration, Environmental Data Service (1968), 80 p.

## ANALYTIC METHODS

During the mechanical (sieve and pipette)-size analysis, the 228 Long Island Sound samples were separated into 13 whole phi-size class intervals ranging from -2 to +10 phi. Pierce and Good's (1966) computer program was used to calculate sample mean size, sorting, skewness, and the sand, silt and clay ratios. The sampels were assigned appropriate textural descriptive terms according to their position within Shepard's (1954) sand, silt, and clay triangular diagram.

Mechanical-size analysis permits every sample to be defined by a point in n-dimensional sample space. In this study, the weight percent in each of the 13 phi-size classes is taken as a unique attribute or variable of the whole sample. With this approach, samples can be compared quantitatively and analyzed in a

# SEDIMENTARY ENVIRONMENTAL ANALYSIS

nonarbitrary manner by means of several multivariate statistical techniques. Cluster analysis, ordination and discriminate analysis were applied to the study area sediment data following the analytical strategy outlined in Figure 2.

Cluster analysis is a technique for grouping samples (Q-mode) or variables (R-mode) that have high similarity indices and then aggregating them at lower levels with other clusters as indicated by their decreasing similarity indices (Sokal and Sneath, 1963; Parks, 1966; Harbaugh and Merriam, 1968). In this manner one can obtain a hierarchical classification that is both objective and unambiguous. As noted by Park (1974), this method seems preferable to other methods of classification which assign arbitrary class boundaries, such as the sand, silt, and clay classifications of Folk (1954) or Shepard (1954).

Sorensen's coefficient (1948), an association similarity index, has been used here in the cluster analyses. This coefficient has been shown to be useful in paleoecologic studies (Park, 1968; Gevirtz, Park, and Friedman, 1971) as well as in sedimentologic studies (Feldhausen, 1970; Feldhausen and Ali, 1974).

Sorensen's coefficient is given by

$$C = \frac{2\sum_{i=1}^{n} \text{Min}(X_{ij}, X_{ik})}{\sum_{i=1}^{n} X_{ij} + \sum_{i=1}^{n} X_{ik}}$$

where X is evaluated for the i-th phi weight percent in the j-th and k-th samples, respectively.

Goodall (1954) described ordination as any number of methods for ordering objects in a uni- or multidimensional continuum. In geology, ordination has been used for obtaining information about the autoecology of fossil species (Park, 1968) and for studying sedimentary depositional environments (Davis, 1970; Feldhausen, 1970).

Ordinations (Q-mode) for this study were constructed by the method of principal components. The method may be visualized as a rigid rotation of the variable axes through the sample space to a new orientation where the variance about each axis is minimized (Davis, 1970). One of these new axes defines a new variable, the principal component, which accounts for the maximum possible variance in the system. Each remaining axis or component accounts for decreasing amounts of the remaining variance. The result of this process is a spatial distribution of the samples in a multidimensional field such that the proximity of one sample to any other sample is inversely proportional to their similarity.

In many problems the nature of the ordination can be deduced by examination of each sample variable or of known environmental parameters with respect to sample position within the ordination. Trend surfaces (Merriam and Harbaugh, 1964; Miesch and Conner, 1968) fitted to sample variables or environmental parameters may be used to quantify attribute gradients over selected planes within the ordination. By comparing several trend surfaces, the environmental significance of the sample clusters, and, hence, the complex of conditions affecting the distribution of the sediments in the natural environment may become apparent.

Discriminant analysis (Harbaugh and Merriam, 1968; Klovan and Billings, 1967; Davies and Ethridge, 1975) enables the investigator to evaluate the statistical significance of a classification and to derive an objective function for identifying unknown samples. In essence, discriminant analysis weights the variables in order to achieve the maximum separation among sample points in previously defined sample groups. These weightings then are used to generate linear equations that can be used to assign unknown samples to their appropriate class.

## APPLICATION TO LONG ISLAND SOUND

### Redundancy Analysis

Redundancy usually is undesirable because it increases the weighting given to an attribute in proportion to the number of variables involved. This tends to bias the analysis toward the environmental aspects represented by the redundant variables. Redundancy can be removed by combining or deleting variables.

R-mode (variable) cluster analysis was used to examine the 13 whole phi-size variables for redundancy. The resulting dendrogram, Figure 3, indicates that none of the variables is clustered with another at an extremely high level of similarity, except perhaps variables 9 and 10 phi. As a further text, two Q-mode (samples) cluster dendrograms were constructed, with and without the 10-phi variable. The results were determined to be essentially identical. Hence, the variable 9 and 10 phi do not seem to bias the study and no variables were deleted or recombined during the remainder of the investigations.

### Sample Classification

The Q-mode cluster dendrogram of the 57 Sound samples (Fig. 3) shows two major sample clusters jointed at a similarity of about C=30 percent. However, they are too broad to be considered meaningful, textural classes. Instead, five sample classes, A through E, were selected from the dendrogram by considering intercluster similarities of about C=70 percent. A distinct band of grain-size curves is associated with each of these clusters (not shown), suggesting that the grouped samples have similar textural properties. Samples within each cluster also reflect environmental similarities, as individual whole-phi variables have environmental significance a priori. Thus, the five clusters may be considered to be sedimentary facies.

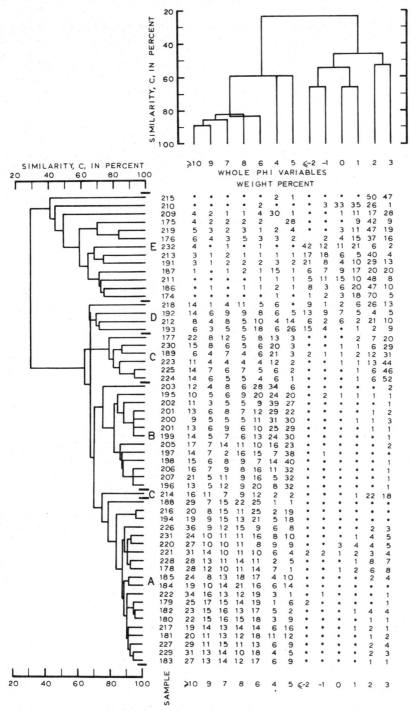

Figure 3. R-mode dendrogram of 13 whole phi-size variables, Q-mode dendrogram partitioned into 5 sedimentary facies, and weight-percent data for Buzas' 57 samples.

The relationship between the R- and Q-mode dendrograms may be visualized by examining the phi weight-percent data presented in Figure 3.

## Interpretation of the Classification

*Ordination* - A three-dimensional, Q-mode ordination (Fig. 4) of the 57 study-area samples was constructed by the method of principal components. The first two principal components account for approximately 66 percent of total variance in the data, whereas the third component only accounts for an additional 15 percent (Table 2). Within this ordination the samples are distributed so that their proximity to one another is inversely proportional to their similarity. Hence gradational relationships among the samples are depicted as well as among the five cluster facies which have been identified in the ordination.

The dendrogram and the ordination have a similar structure, although one depicts discrete relationships and the other gradational relationships. As in the dendrogram, the silty samples of Facies A and B are grouped together in the ordination with high intragroup similarities, whereas the sandy samples, Facies E, are widely separated.

*Gradient Analysis* - Preliminary environmental inferences may be drawn from Table 2. The first principal component, which accounts for about one-half of the variance within the data, represents the percent sand gradient from left to right across Figure 4. Only two samples contain less sand than Sample 207. Sample 174 has the highest proportion of sand-sized particles. The gradient of mean diameter is similar to that of percent sand. These two textural gradients suggest a marked increase in environmental energy from Sample 207 to Sample 174. Textural gradients along the second and third components are ill-defined, because they each account for only about 15 percent of the total variance.

A more precise picture of the environmental structure underlying the ordination may be obtained from the interpretation of Figure 5. Here, second-degree trend surfaces for the phi mean diameter, sorting and skewness have been contoured. The statistics for these surfaces, and for other surfaces which are not plotted, are reported in Table 3.

The mean diameter increases in a gentle curved arc from Sample 207 to Sample 174. The most rapid changes occur within Facies A and B. Sorting takes a domal shape, with the poorest sorting (highest phi value) located to the left of the ordination's center. The sorting trend surface does not fit the data as well as the mean diameter surface, but it provides a better fit than the skewness surface which explains only 45 percent of the variation in the data.

Not shown in Figure 5 are trend surfaces for the percent sand and for water depth, although their statistics are listed in Table 3. As might be expected, the first- and second-degree surfaces

# SEDIMENTARY ENVIRONMENTAL ANALYSIS 83

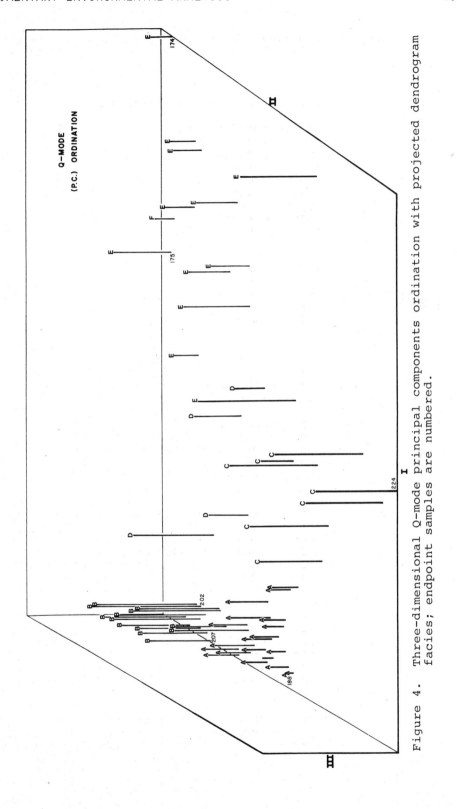

Figure 4. Three-dimensional Q-mode principal components ordination with projected dendrogram facies; endpoint samples are numbered.

Table 2. Ordination of Buzas' sediment data.

A. Axis Data

| Parameter | Principal Components Ordination Axes | | |
|---|---|---|---|
| | I (X) | II (Y) | III (Z) |
| Contribution per Eigenvalue | 49.44 (%) | 16.42 (%) | 14.77 (%) |
| End-Point Samples | 207,174 | 224,175 | 188,202 |
| Axis Length (arbitrary units) | 78.90 | 58.60 | 49.00 |

B. End-Point Data

| End Point | Parameter | | | | |
|---|---|---|---|---|---|
| | Phi Mean Diameter | Phi Sorting | Phi Skewness | Percent Sand | Facies |
| 207 | 6.77 | 2.78 | 0.44 | 5.4 | B |
| 174 | 1.21 | 0.50 | -0.20 | 99.4 | E |
| 224 | 4.69 | 3.34 | 0.54 | 65.2 | C |
| 175 | 3.21 | 1.80 | 0.61 | 59.1 | D |
| 188 | 7.98 | 2.44 | 0.21 | 1.0 | A |
| 202 | 5.33 | 2.55 | 0.65 | 39.9 | B |

for percent sand provide essentially the same information as those for the mean diameter. The corresponding water-depth surfaces are not significant, and, hence, were not presented.

At Facies A, shown on Figure 5, the coincidence of high phi mean diameters, poor sorting, and near zero skewness is indicative of a quiescent environment characterized by week currents and gravitational settling. A deep-water depositional environment might be expected. However, relatively shallow water is indicated by the general absence of planktonic Foraminifera and by the species of benthonic Foraminifera observed at these sample locations (see Buzas, 1965). Facies B, which lies in close proximity to Facies A, has textural properties not unlike those of its neighbor. Facies C and D have properties intermediate between those of A and B and of E. This portion of the ordination is associated with a decreasing mean diameter gradient, a broad phi sorting high and moderately positive skewness. Much stronger current processing is suggested by the coarse texture of the samples composing Facies E. However, its moderately poor sorting precludes prolonged winnowing such as that experienced by beach sands. Instead, tidal currents and storm-wave action are postulated to provide the depositional environment of this facies (Greenwood, 1969; Allen, 1971; Jones, 1971).

## Testing and Verification

A Wilks' lambda analysis of dispersion using Rao's (1968) program was performed to test the statistical significance of the

# SEDIMENTARY ENVIRONMENTAL ANALYSIS

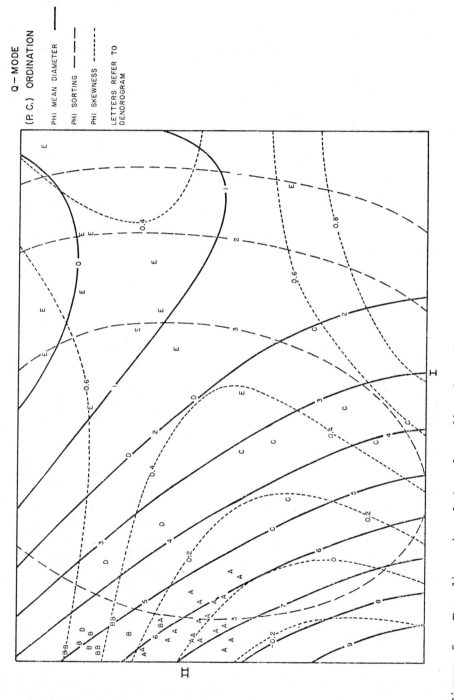

Figure 5. Two-dimensional Q-mode ordination with second-degree trend surfaces, for phi mean diameter, sorting, and skewness.

Table 3. Ordination trend-surface statistics.

| Parameter | Statistic | Surface | |
|---|---|---|---|
| | | First Degree | Second Degree |
| Phi Mean Dia. | Std. Dev. | 0.93 | 0.68 |
| | Variation explained (%) | 85.0 | 92.0 |
| | Correlation Coef. | 0.92 | 0.96 |
| Phi Sorting | Std. Dev. | 0.65 | 0.44 |
| | Variation explained (%) | 16.0 | 62.0 |
| | Correlation Coef. | 0.40 | 0.78 |
| Phi Skewness | Std. Dev. | 0.32 | 0.26 |
| | Variation explained (%) | 17.0 | 45.0 |
| | Correlation Coef. | 0.42 | 0.67 |
| Percent Sand | Std. Dev. | 1.09 | 0.86 |
| | Variation explained (%) | 89.0 | 93.0 |
| | Correlation Coef. | 0.94 | 0.96 |
| Water Depth | Std. Dev. | 10.67 | 10.03 |
| | Variation explained (%) | 5.0 | 16.0 |
| | Correlation Coef. | 0.21 | 0.40 |

dendrogram classification. The analysis calculated low values of lambda and high values of the F-test statistic. Thus, the dendrogram classification exhibits low dispersion. The classification also may be considered satisfactory, because the F-test statistic showed the variable means for the five facies to be unequal (statistically different).

Dixon's (1970) discriminant analysis program BMD07M was used to evaluate further the classification. It also demonstrated the inequality of the facies' multivariate means.

The classification and ordination can be verified by evaluating their consistency with field relations and known or inferred environmental processes. Figures 4 and 5 show the correspondence between the five cluster facies and their projection in the ordination; Figure 1 shows the field relations of the five facies.

The distribution of the five sedimentary facies identified by Q-mode cluster analysis is consistent with the observations of other investigators (McCrone, Ellis, and Charmatz, 1961; Grim, Drake, and Heirtzler, 1970). Facies A and B, which are composed chiefly of clayey silts and sandy silts, are the most abundant in the deeper parts of Long Island Sound, particularly in the Central and Western Basins, where both tidal and nontidal currents are weak and wave action does not penetrate deep enough to disturb and rework the sediments. The bottom-drifter data, discussed earlier, leads to the conclusion that fine, river-borne sediment is likely to move generally westward, remaining close to the

Connecticut shore. This conclusion is supported by many of the northern bottom samples (Fig. 1). Soft, fine-grained sediments cover about one-third of the Sound. In many patchy locations their gas content is so high that they obscure seismic reflections from the underlying strata (Grim, Drake, and Heirtzler, 1970). Relatively few organisms can live on soft bottoms, which may account for the scarcity of food organisms in these basins.

Coarse sands commonly form the beaches and harbor sediments along the shores of the Sound. These deposits extend to the depths of about 10 m. The sands of Facies E occur on the large shoal areas and sills (Samples 191) and in areas of current bedding and megaripples (Sample 215). Strong currents at these locations, such as those measured near Six Mile Reef, prevent deposition of fine-grained sediments and leave a cover of coarser sediment.

The medium to coarse size and poor sorting (Fig. 5) of the Facies E sands suggest a predominance of current rather than wave processing. The general lack of prolonged winnowing by wave action is not surprising if the limited fetch of Long Island Sound (Fig. 1) and the average wind conditions (Table 1) are considered. As indicated in the description of the study area, average wave base is estimated to be about 6 m, whereas that from infrequent storms is estimated to be about 25 to 30 m.

The sands of Facies E are of economic importance; much of the Sound's shellfish production is taken from this facies. This facies also may be valuable as potential sources of sand and gravel for the construction industry.

Facies C and D are associated with environments whose energies are less than that for Facies E but greater than for Facies A and B. Facies C samples were taken from the southern flank of the Western and Central Basins and the deeper portions of Mattituck Sill. Facies D sample 212 was also taken from deep water on this sill, whereas other samples were in the vicinity of the Connecticut shore.

## Identification of New Samples

The dendrogram classification (Fig. 3) was extended with the aid of discriminant analysis (Dixon, 1970, program MBD07M) to include 171 additional bottom samples obtained while the junior author was associated with the Marine Sciences Research Center, SUNY Stony Brook, New York. Output from discriminant analysis was compared with the original data and the results of Q-mode ordinations (not shown) to ensure that the classification of the new samples is consistent with the original classification.

All of the 13 whole phi variables were employed in the stepwise discriminant analysis except for the 0 phi size. After 12 steps, this variable failed to exhibit a F-level significant for further computation. From Table 4 it may be seen that the 8, 5, 10, 6, and 4 phi sizes have the highest F- and U-statistic values, and hence contributed most to the discrimination of the 5 facies.

Table 4. Variables and F-values for stepwise discrimination.

| Discriminant Step | Variable Entered (Phi) | F-Value To Enter | U-Statistic | Number Of Variables Included |
|---|---|---|---|---|
| 1 | 8 | 130.82 | 0.2988 | 1 |
| 2 | 5 | 81.71 | 0.1209 | 2 |
| 3 | >10 | 34.10 | 0.0747 | 3 |
| 4 | 6 | 29.31 | 0.0488 | 4 |
| 5 | 4 | 17.60 | 0.0369 | 5 |
| 6 | 7 | 16.39 | 0.0284 | 6 |
| 7 | 9 | 9.38 | 0.0242 | 7 |
| 8 | 3 | 2.30 | 0.0232 | 8 |
| 9 | -1 | 1.52 | 0.0226 | 9 |
| 10 | <-2 | 0.51 | 0.0223 | 10 |
| 11 | 1 | 0.22 | 0.0223 | 11 |
| 12 | 2 | 0.09 | 0.0222 | 12 |

Figure 6 is a canonical variate (Reyment and Ramden, 1970) scatterdiagram of the 228 samples produced as an option during the discriminant analysis. As in principal components analysis, the X-axis is inclined in the direction of greatest variability between the means of the samples. The Y-axis is normal to the X-axis and is inclined in the direction of the next greatest variability. The X-axis for Figure 6 accounts for almost 92 percent of the total dispersion, whereas the Y-axis accounts for the remaining 8 percent.

In this scatterdiagram there seems to be little facies overlap except between Facies C and D. Overlap of these facies could be minimized if border line samples were reclassified. However, the classification of the samples provided in Table 5 and reflected in Table 6 and Figures 6 and 7 is believed to be consistent with the basic data and grain-size curves.

The inequality of the facies' variable means is shown in Figure 6, as well as Table 6. This table also shows that the facies' average grain-size statistics are distinct.

## Discussion of Facies Chart

A classification of 228 Long Island Sound bottom samples, established with the strategy diagrammed in Figure 2, is presented in Table 5 and as the facies chart, Figure 7. In addition to listing the facies associated with each sample, Table 5 also lists sample phi mean diameter, plus sorting, percent sand size, and sampling location.

The distribution of the five sedimentary facies shown in Figure 7 is consistant with the interpretation of the dendrogram (Fig. 2) and ordination (Figs. 3 and 4) discussed previously, and it seems to be consistent with the distribution of sediments within the Eastern Basin observed by Grim, Drake and, Heirtzler (1970)

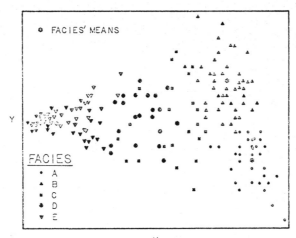

Figure 6. Canonical scatterdiagram of 5 dendrogram-discriminant facies. The facies' multivariate means are separated in sample space; minor overlap of Facies C and D are shown.

and by Akpati (1974). The overall sedimentary patterns are not unlike those discerned by M.G. Gross (1974, personal communication) based on silt, fine-sand, and coarse-sand distributions.

Four major environments of deposition are suggested by Figure 7. The first is the nearshore zone where sands are usually derived from local sources, such as the erosion of bluffs, and other beaches, and they are reworked and transported by waves with an average wave base of less than 6 m. Long Island Sound generally is well protected from the large waves generated in the North Atlantic. However, infrequent severe storms may generate waves as high as 3 m with a wave base that extends to 25 to 30 m.

In the second depositional environment, waves from severe storms may process the deep-water, coarse-grained sediments (Facies E and D) which predominate in the Eastern Basin. However, tidal currents are the dominant processing agent. In the Eastern Basin and near Six Mile Reef, these currents are strong enough to form megaripples and to remove the finer grained fraction of the sediment deposits.

Shoal areas comprise the third depositional environment. Deposits of Facies E sands lie on top of the north-south sills and on the large submerged banks, such as those around Stratford Shoals or Cable and Anchor Reef (Fig. 1). Here strong tidal currents prevent the deposition of fine-grained sediments and usually leave a cover of coarse sand and gravel. Facies D and C sediments may occur in the deeper portions or on the flanks of the shoals and sills.

Table 5. Location and classification Long Island Sound bottom samples.

| SAMPLE NO. | NORTH LAT. DEG. | MIN. | WEST LONG. DEG. | MIN. | PHI MEAN DIA. | PHI SORTING | PERCENT SAND | FACIES |
|---|---|---|---|---|---|---|---|---|
| 1 | 40 | 58.5 | 73 | 20.0 | 6.36 | 2.38 | 15.3 | A |
| 2 | 40 | 58.5 | 73 | 19.0 | 5.05 | 2.38 | 22.9 | B |
| 3 | 40 | 58.0 | 73 | 17.5 | 6.49 | 2.11 | 22.9 | A |
| 4 | 40 | 58.0 | 73 | 11.0 | 2.132 | .695 | 98.6 | E |
| 5 | 40 | 56.5 | 73 | 10.5 | 5.35 | 1.90 | 29.4 | B |
| 6 | 40 | 56.0 | 73 | 9.5 | 4.24 | 1.80 | 75.5 | D |
| 7 | 40 | 55.5 | 73 | 12.0 | .413 | 1.03 | 97.7 | E |
| 8 | 40 | 56.5 | 73 | 15.5 | 1.83 | 2.77 | 83.5 | E |
| 9 | 40 | 56.5 | 73 | 17.5 | .919 | .910 | 94.2 | E |
| 10 | 40 | 57.0 | 73 | 19.2 | 3.34 | 3.04 | 69.1 | D |
| 11 | 40 | 57.5 | 73 | 20.5 | 5.99 | 2.80 | 22.9 | A |
| 12 | 40 | 57.5 | 73 | 21.5 | 6.02 | 2.40 | 32.1 | B |
| 13 | 40 | 57.0 | 73 | 27.5 | .543 | 1.176 | 95.3 | E |
| 14 | 40 | 58.0 | 73 | 28.0 | 2.63 | 2.39 | 79.4 | E |
| 15 | 40 | 59.5 | 73 | 28.5 | 4.51 | 2.11 | 55.3 | C |
| 16 | 41 | 0.5 | 73 | 28.0 | 5.14 | 2.38 | 29.4 | C |
| 17 | 41 | 1.5 | 73 | 27.0 | 6.29 | 2.13 | 17.7 | A |
| 18 | 41 | 1.0 | 73 | 25.0 | 1.38 | 1.06 | 95.0 | E |
| 19 | 41 | 0.5 | 73 | 25.5 | 6.77 | 2.34 | 17.4 | A |
| 20 | 41 | 0.0 | 73 | 26.0 | 1.73 | 1.95 | 89.0 | E |
| 21 | 40 | 59.0 | 73 | 25.5 | 3.52 | 2.93 | 69.0 | D |
| 22 | 40 | 59.5 | 73 | 24.0 | 2.94 | 2.54 | 76.1 | E |
| 23 | 40 | 59.5 | 73 | 23.0 | 1.50 | 1.23 | 90.1 | E |
| 24 | 40 | 59.5 | 73 | 21.0 | 4.79 | 2.91 | 42.6 | C |
| 25 | 40 | 59.5 | 73 | 19.5 | 5.79 | 2.34 | 35.7 | B |
| 26 | 40 | 59.5 | 73 | 17.5 | 4.33 | 2.74 | 50.9 | D |
| 27 | 40 | 58.2 | 73 | 10.4 | 5.45 | 2.14 | 35.0 | C |
| 28 | 40 | 57.8 | 73 | 9.9 | .379 | .926 | 96.1 | E |
| 29 | 40 | 58.9 | 73 | 9.2 | 2.95 | 2.63 | 79.6 | E |
| 30 | 41 | 0.2 | 73 | 9.0 | 1.96 | 1.000 | 91.2 | E |
| 31 | 41 | 2.0 | 73 | 6.0 | 1.02 | .897 | 94.3 | E |
| 32 | 41 | 3.0 | 73 | 5.4 | .967 | 1.86 | 93.1 | E |
| 33 | 41 | 4.3 | 73 | 5.5 | .236 | .850 | 98.9 | E |
| 34 | 41 | 4.0 | 73 | 5.2 | 2.71 | 3.20 | 83.8 | E |
| 35 | 41 | 2.3 | 73 | 10.9 | 6.40 | 2.33 | 19.1 | A |
| 36 | 41 | 0.5 | 73 | 15.0 | 6.49 | 2.54 | 18.0 | A |
| 37 | 40 | 56.4 | 73 | 20.8 | 3.11 | 2.60 | 78.2 | E |
| 38 | 40 | 56.3 | 73 | 20.7 | 2.18 | .728 | 92.2 | E |
| 39 | 40 | 56.2 | 73 | 20.7 | 1.90 | .684 | 95.5 | E |
| 40 | 40 | 55.8 | 73 | 20.6 | 1.91 | .791 | 95.6 | E |
| 41 | 40 | 56.7 | 73 | 20.6 | 2.29 | .593 | 100.0 | E |
| 42 | 40 | 55.6 | 73 | 20.6 | 1.86 | .779 | 100.0 | E |
| 43 | 40 | 55.7 | 73 | 20.7 | 1.84 | .611 | 100.0 | E |
| 44 | 40 | 56.5 | 73 | 20.6 | 1.10 | .731 | 96.4 | E |
| 45 | 40 | 56.6 | 73 | 20.6 | 2.73 | 2.04 | 80.7 | E |
| 46 | 40 | 55.7 | 73 | 20.5 | 1.45 | .726 | 100.0 | E |
| 47 | 40 | 55.8 | 73 | 20.5 | 1.40 | .741 | 100.0 | E |
| 48 | 40 | 56.3 | 73 | 20.5 | 1.46 | 1.00 | 100.0 | E |
| 49 | 40 | 56.4 | 73 | 20.6 | .99 | .706 | 100.0 | E |
| 50 | 40 | 56.6 | 73 | 20.6 | .94 | .826 | 100.0 | E |
| 51 | 40 | 56.9 | 73 | 20.7 | 2.48 | 2.99 | 75.3 | E |
| 52 | 41 | 9.5 | 72 | 25.0 | .224 | .691 | 100.0 | E |
| 53 | 41 | 13.0 | 72 | 22.5 | 4.25 | 3.65 | 48.3 | D |
| 54 | 41 | 14.7 | 72 | 19.5 | 1.58 | .417 | 100.0 | E |
| 55 | 41 | 16.3 | 72 | 20.5 | 3.83 | 1.85 | 65.6 | D |
| 56 | 41 | 12.5 | 72 | 38.5 | 4.04 | 2.30 | 70.4 | D |
| 57 | 41 | 9.5 | 72 | 47.5 | 4.72 | 2.06 | 45.1 | C |
| 58 | 41 | 9.4 | 73 | 5.3 | 1.99 | .680 | 100.0 | E |
| 59 | 40 | 55.9 | 73 | 18.5 | 1.52 | .354 | 99.4 | E |
| 60 | 40 | 56.1 | 73 | 18.5 | 1.63 | .441 | 98.2 | E |
| 61 | 40 | 56.3 | 73 | 18.4 | 1.00 | .780 | 98.1 | E |
| 62 | 40 | 56.5 | 73 | 18.4 | 1.25 | .559 | 98.3 | E |
| 63 | 40 | 56.7 | 73 | 18.4 | 1.17 | .617 | 98.6 | E |
| 64 | 40 | 56.9 | 73 | 18.3 | 1.36 | .454 | 98.3 | E |
| 65 | 40 | 57.1 | 73 | 18.3 | 1.08 | .691 | 97.1 | E |
| 66 | 40 | 57.3 | 73 | 18.3 | 3.34 | 2.49 | 64.6 | D |
| 67 | 41 | 0.2 | 73 | 26.2 | 1.76 | 1.102 | 89.1 | E |
| 68 | 40 | 58.1 | 73 | 25.0 | 2.27 | .582 | 97.9 | E |
| 69 | 40 | 57.8 | 73 | 26.6 | 5.36 | 2.32 | 29.0 | B |
| 70 | 40 | 58.5 | 73 | 28.4 | 3.72 | 2.70 | 49.3 | D |
| 71 | 40 | 59.4 | 73 | 28.7 | 6.20 | 1.25 | 1.4 | B |
| 72 | 40 | 59.1 | 73 | 28.1 | 6.18 | 1.79 | 9.6 | B |
| 73 | 40 | 58.9 | 73 | 27.0 | 6.26 | 1.76 | 7.5 | B |
| 74 | 41 | 0.1 | 73 | 26.2 | 3.95 | 3.10 | 48.6 | D |
| 75 | 40 | 59.6 | 73 | 25.9 | 6.26 | 1.77 | 11.9 | A |
| 76 | 40 | 59.8 | 73 | 24.7 | 1.27 | 2.42 | 86.1 | E |
| 77 | 41 | 0.0 | 73 | 25.2 | - .891 | 1.042 | 98.5 | E |
| 78 | 41 | 0.4 | 73 | 25.7 | 5.32 | 2.81 | 37.9 | C |
| 79 | 41 | 0.3 | 73 | 27.2 | 5.16 | 1.70 | 24.7 | B |
| 80 | 40 | 59.4 | 73 | 27.0 | 5.46 | 2.24 | 19.0 | B |
| 81 | 40 | 59.7 | 73 | 28.6 | 5.70 | .83 | 4.0 | B |
| 82 | 40 | 59.9 | 73 | 30.3 | 6.30 | 1.28 | 6.6 | B |

# SEDIMENTARY ENVIRONMENTAL ANALYSIS

| SAMPLE NO. | NORTH LAT. DEG. | MIN. | WEST LONG. DEG. | MIN. | PHI MEAN DIA. | PHI SORTING | PERCENT SAND | FACIES |
|---|---|---|---|---|---|---|---|---|
| 83 | 41 | 0.9 | 73 | 29.1 | 5.60 | 1.34 | 8.5 | B |
| 84 | 41 | 1.1 | 73 | 26.9 | 4.32 | 1.98 | 28.6 | B |
| 85 | 41 | 2.3 | 73 | 23.1 | 2.70 | 1.160 | 87.0 | E |
| 86 | 41 | 1.6 | 73 | 20.2 | 4.12 | 1.84 | 33.3 | C |
| 87 | 41 | 3.4 | 73 | 18.7 | 2.59 | 2.23 | 71.9 | E |
| 88 | 41 | 3.4 | 73 | 17.9 | 5.78 | 1.60 | 9.8 | B |
| 89 | 41 | 4.6 | 73 | 19.6 | 5.26 | 1.15 | 12.4 | B |
| 90 | 41 | 4.1 | 73 | 12.5 | 3.58 | 2.07 | 59.8 | D |
| 91 | 40 | 57.5 | 73 | 4.8 | 5.97 | 2.04 | 17.4 | B |
| 92 | 41 | 0.5 | 72 | 47.0 | 6.11 | 1.73 | 7.2 | B |
| 93 | 41 | 5.1 | 72 | 30.6 | 1.58 | 0.400 | 98.4 | E |
| 94 | 41 | 11.0 | 72 | 22.0 | 1.12 | .652 | 99.1 | E |
| 95 | 41 | 12.7 | 72 | 21.0 | .328 | 1.303 | 95.5 | E |
| 96 | 41 | 10.7 | 72 | 13.1 | -0.654 | 1.340 | 99.0 | E |
| 97 | 41 | 18.0 | 72 | 4.8 | 4.25 | 0.74 | 29.7 | B |
| 98 | 41 | 17.2 | 72 | 4.4 | 2.27 | 1.060 | 92.8 | E |
| 99 | 41 | 16.6 | 72 | 3.0 | 4.34 | 1.29 | 48.7 | C |
| 100 | 41 | 16.2 | 72 | 4.4 | 3.05 | .784 | 89.7 | E |
| 101 | 41 | 14.0 | 72 | 7.0 | .671 | 2.840 | 80.4 | E |
| 102 | 41 | 0.9 | 72 | 41.4 | 4.55 | 0.83 | 19.8 | B |
| 103 | 41 | 10.4 | 72 | 40.3 | 3.35 | .946 | 81.6 | E |
| 104 | 41 | 13.6 | 72 | 40.0 | 2.66 | 1.13 | 89.3 | E |
| 105 | 41 | 13.4 | 72 | 40.3 | .78 | .950 | 99.5 | E |
| 106 | 41 | 13.4 | 72 | 54.6 | 1.02 | 1.55 | 90.2 | E |
| 107 | 41 | 1.4 | 72 | 53.7 | 5.14 | 1.80 | 22.5 | B |
| 108 | 41 | 0.0 | 73 | 5.0 | 2.19 | 2.49 | 76.4 | E |
| 109 | 41 | 9.5 | 73 | 5.4 | 5.39 | 1.934 | 28.2 | B |
| 110 | 40 | 59.3 | 73 | 30.0 | 5.81 | 3.29 | 27.2 | A |
| 111 | 40 | 57.3 | 73 | 34.4 | 7.28 | 2.47 | 10.4 | A |
| 112 | 40 | 55.5 | 73 | 41.0 | 7.86 | 1.69 | .5 | A |
| 113 | 40 | 54.3 | 73 | 39.5 | 3.52 | 3.06 | 77.1 | E |
| 114 | 40 | 52.3 | 73 | 44.3 | 1.80 | 2.63 | 77.1 | E |
| 115 | 40 | 52.6 | 73 | 45.0 | 5.64 | 3.05 | 33.3 | C |
| 116 | 40 | 50.5 | 73 | 45.8 | 5.88 | 3.76 | 39.0 | C |
| 117 | 40 | 48.0 | 73 | 47.3 | 3.23 | 2.55 | 78.1 | E |
| 118 | 40 | 48.4 | 73 | 49.2 | 2.86 | 1.69 | 82.2 | E |
| 119 | 40 | 47.6 | 73 | 45.9 | 5.00 | 0.87 | 12.6 | B |
| 120 | 40 | 49.2 | 73 | 46.8 | 5.20 | 1.63 | 26.5 | B |
| 121 | 40 | 50.1 | 73 | 47.1 | 6.77 | 2.05 | 3.3 | A |
| 122 | 40 | 51.5 | 73 | 39.5 | 2.01 | .734 | 97.4 | E |
| 123 | 40 | 51.7 | 73 | 39.6 | 6.49 | 2.18 | 10.9 | A |
| 124 | 40 | 52.2 | 73 | 39.9 | 4.82 | 2.20 | 38.3 | B |
| 125 | 40 | 52.0 | 73 | 40.2 | 3.80 | 2.76 | 57.1 | D |
| 126 | 40 | 51.8 | 73 | 40.5 | 5.56 | 2.03 | 22.0 | B |
| 127 | 40 | 51.6 | 73 | 40.1 | | | | |
| 128 | 40 | 51.6 | 73 | 40.6 | 1.71 | .592 | 99.7 | E |
| 129 | 40 | 51.4 | 73 | 40.3 | 1.55 | .443 | 99.9 | E |
| 130 | 40 | 51.1 | 73 | 39.9 | .202 | 1.521 | 97.7 | E |
| 131 | 40 | 50.8 | 73 | 39.8 | .091 | 1.896 | 93.4 | E |
| 132 | 40 | 50.8 | 73 | 39.4 | .905 | 1.03 | 99.2 | E |
| 133 | 40 | 50.5 | 73 | 39.8 | .109 | 1.646 | 96.7 | E |
| 134 | 40 | 50.4 | 73 | 39.6 | 1.59 | 2.95 | 84.4 | E |
| 135 | 40 | 50.4 | 73 | 39.4 | 1.60 | 1.13 | 99.4 | E |
| 136 | 40 | 50.3 | 73 | 39.4 | 2.78 | 1.88 | 80.2 | E |
| 137 | 40 | 50.1 | 73 | 39.4 | .210 | 1.733 | 94.0 | E |
| 138 | 40 | 47.7 | 73 | 46.1 | 5.56 | 2.03 | 22.0 | B |
| 139 | 40 | 47.9 | 73 | 45.7 | 4.46 | 2.88 | 40.0 | C |
| 140 | 40 | 48.1 | 73 | 46.4 | 5.51 | 3.08 | 39.7 | C |
| 141 | 40 | 48.5 | 73 | 46.1 | 4.34 | 2.74 | 51.4 | D |
| 142 | 40 | 48.6 | 73 | 46.0 | 1.66 | 1.62 | 93.7 | E |
| 143 | 40 | 49.7 | 73 | 43.2 | 6.55 | 1.95 | 8.8 | A |
| 144 | 40 | 49.9 | 73 | 43.1 | 7.12 | 2.00 | 4.6 | A |
| 145 | 40 | 49.7 | 73 | 44.0 | .506 | 1.019 | 99.8 | E |
| 146 | 40 | 50.2 | 73 | 43.9 | 2.29 | 3.51 | 70.1 | E |
| 147 | 40 | 50.2 | 73 | 43.9 | 3.12 | 3.53 | 58.1 | D |
| 148 | 40 | 50.1 | 73 | 44.8 | 5.41 | 2.43 | 30.1 | B |
| 149 | 40 | 50.4 | 73 | 44.3 | 6.47 | 2.09 | 11.0 | B |
| 150 | 40 | 50.7 | 73 | 44.8 | 5.20 | 2.40 | 27.2 | B |
| 151 | 40 | 50.9 | 73 | 44.5 | 3.38 | 2.98 | 67.1 | D |
| 152 | 40 | 51.4 | 73 | 44.0 | 1.25 | .772 | 99.4 | E |
| 153 | 41 | 18.0 | 72 | 5.4 | 2.81 | 1.81 | 77.3 | E |
| 154 | 41 | 16.2 | 72 | 10.7 | | | | |
| 155 | 41 | 15.2 | 72 | 20.6 | 1.47 | 1.050 | 98.2 | E |
| 156 | 41 | 14.4 | 72 | 28.8 | 5.52 | 1.09 | 2.3 | B |
| 157 | 41 | 9.8 | 72 | 29.8 | 1.57 | .393 | 95.7 | E |
| 158 | 41 | 7.3 | 72 | 29.1 | 1.54 | .447 | 97.4 | E |
| 159 | 41 | 5.1 | 72 | 28.7 | 2.48 | .555 | 95.7 | E |
| 160 | 41 | 1.4 | 72 | 46.1 | 6.23 | 2.44 | 18.4 | B |
| 161 | 41 | 8.9 | 72 | 51.9 | 6.22 | 1.71 | 1.9 | B |
| 162 | 41 | 11.4 | 72 | 53.1 | 6.22 | 2.06 | 13.5 | B |
| 163 | 41 | 11.2 | 72 | 58.6 | 6.67 | 2.51 | 13.4 | B |
| 164 | 41 | 6.7 | 73 | 1.5 | 6.41 | 1.95 | 10.4 | A |
| 165 | 41 | 1.7 | 73 | 1.5 | 5.11 | 2.54 | 35.1 | C |
| 166 | 41 | 6.3 | 72 | 52.8 | 5.89 | 1.88 | 16.8 | B |
| 167 | 41 | 9.4 | 72 | 52.4 | 6.10 | 1.86 | 8.4 | B |
| 168 | 41 | 12.1 | 72 | 53.6 | 4.72 | 2.04 | 36.6 | C |
| 169 | 41 | 13.4 | 72 | 54.6 | .540 | 1.235 | 98.6 | E |
| 170 | 41 | 14.2 | 72 | 55.0 | .586 | 1.163 | 97.9 | E |
| 171 | 41 | 9.0 | 72 | 53.3 | 6.07 | 1.62 | 6.7 | B |

| SAMPLE NO. | NORTH LAT. DEG. | MIN. | WEST LONG. DEG. | MIN. | PHI MEAN DIA. | PHI SORTING | PERCENT SAND | FACIES |
|---|---|---|---|---|---|---|---|---|
| 172 | 40 | 56.8 | 73 | 24.9 | 5.21 | 2.52 | 37.2 | C |
| 173 | 40 | 58.8 | 73 | 28.3 | 3.40 | 2.33 | 69.7 | D |
| 174 | 41 | 0.0 | 72 | 58.0 | 1.21 | 0.500 | 99.4 | E |
| 175 | 41 | 1.1 | 72 | 58.4 | 3.21 | 1.80 | 59.1 | D |
| 176 | 41 | 2.0 | 72 | 58.6 | 3.00 | 3.00 | 77.1 | E |
| 177 | 41 | 2.9 | 72 | 58.9 | 6.08 | 3.58 | 41.4 | C |
| 178 | 41 | 3.8 | 72 | 59.2 | 7.09 | 3.43 | 23.9 | A |
| 179 | 41 | 4.6 | 72 | 59.5 | 7.61 | 2.91 | 3.9 | A |
| 180 | 41 | 5.5 | 72 | 59.7 | 7.49 | 2.59 | 5.4 | A |
| 181 | 41 | 6.4 | 73 | 0.0 | 6.92 | 2.75 | 14.3 | A |
| 182 | 41 | 7.2 | 73 | 0.2 | 7.33 | 2.90 | 13.2 | A |
| 183 | 41 | 8.1 | 73 | 0.5 | 7.56 | 2.81 | 8.1 | A |
| 184 | 41 | 9.0 | 73 | 0.8 | 7.17 | 2.53 | 7.0 | A |
| 185 | 41 | 9.8 | 73 | 1.1 | 7.22 | 2.84 | 10.9 | A |
| 186 | 41 | 10.7 | 73 | 1.3 | 1.05 | 1.10 | 95.4 | E |
| 187 | 40 | 54.5 | 73 | 37.7 | 1.44 | 2.46 | 95.2 | E |
| 188 | 40 | 55.1 | 73 | 42.2 | 7.98 | 2.44 | 1.0 | A |
| 189 | 41 | 6.9 | 73 | 10.1 | 3.89 | 3.00 | 69.5 | C |
| 191 | 41 | 15.2 | 72 | 55.0 | 0.943 | 3.35 | 88.1 | E |
| 192 | 41 | 13.8 | 72 | 54.0 | 3.80 | 4.82 | 49.8 | D |
| 193 | 41 | 11.5 | 73 | 1.6 | 3.58 | 3.93 | 38.2 | D |
| 194 | 41 | 12.2 | 72 | 45.9 | 7.06 | 2.54 | 4.6 | A |
| 195 | 41 | 11.5 | 72 | 45.6 | 5.40 | 2.70 | 30.6 | B |
| 196 | 41 | 10.7 | 72 | 45.4 | 6.23 | 2.50 | 9.1 | B |
| 197 | 41 | 9.8 | 72 | 45.1 | 6.15 | 2.58 | 8.3 | B |
| 198 | 41 | 9.1 | 72 | 44.8 | 6.11 | 2.65 | 15.1 | B |
| 199 | 41 | 8.2 | 72 | 44.6 | 5.85 | 2.65 | 25.0 | B |
| 200 | 41 | 7.4 | 72 | 44.3 | 5.40 | 2.42 | 32.9 | B |
| 201 | 41 | 6.6 | 72 | 44.0 | 5.70 | 2.65 | 22.6 | B |
| 202 | 41 | 5.8 | 72 | 43.8 | 5.33 | 2.55 | 39.9 | B |
| 203 | 41 | 4.2 | 72 | 43.3 | 5.73 | 2.57 | 35.8 | B |
| 204 | 41 | 3.4 | 72 | 43.0 | 5.85 | 2.66 | 26.4 | B |
| 205 | 41 | 2.5 | 72 | 42.8 | 6.43 | 2.76 | 18.3 | B |
| 206 | 41 | 1.7 | 72 | 42.5 | 6.43 | 2.68 | 11.4 | B |
| 207 | 41 | 0.9 | 72 | 42.2 | 6.77 | 2.78 | 5.4 | B |
| 209 | 41 | 1.5 | 72 | 35.0 | 3.08 | 2.34 | 87.6 | E |
| 210 | 41 | 5.5 | 72 | 27.0 | 0.480 | 1.130 | 99.7 | E |
| 211 | 41 | 7.6 | 72 | 27.5 | 0.613 | 1.40 | 99.7 | E |
| 212 | 41 | 8.4 | 72 | 27.7 | 3.84 | 3.72 | 50.4 | D |
| 213 | 41 | 9.1 | 72 | 27.9 | 0.539 | 2.00 | 91.6 | E |
| 214 | 41 | 9.8 | 72 | 28.1 | 5.51 | 3.56 | 42.8 | C |
| 215 | 41 | 10.5 | 72 | 28.3 | 2.05 | 0.586 | 99.3 | E |
| 216 | 41 | 15.2 | 72 | 28.5 | 7.08 | 2.51 | 2.0 | A |
| 217 | 40 | 55.0 | 73 | 26.6 | 6.97 | 2.79 | 10.8 | A |
| 218 | 40 | 59.0 | 73 | 29.9 | 3.85 | 3.90 | 63.3 | D |
| 219 | 40 | 57.9 | 73 | 29.5 | 2.40 | 1.800 | 81.7 | E |
| 220 | 40 | 57.6 | 73 | 26.8 | 6.79 | 3.59 | 24.5 | A |
| 221 | 40 | 57.2 | 73 | 21.2 | 7.12 | 3.83 | 20.4 | A |
| 222 | 40 | 57.0 | 73 | 10.1 | 8.19 | 2.75 | 4.9 | A |
| 223 | 40 | 59.1 | 73 | 10.7 | 4.18 | 2.80 | 70.9 | A |
| 224 | 41 | 0.0 | 73 | 11.0 | 4.69 | 3.34 | 65.2 | C |
| 225 | 41 | 0.8 | 73 | 11.2 | 4.91 | 3.37 | 59.3 | C |
| 226 | 41 | 1.7 | 73 | 11.5 | 7.99 | 3.05 | 10.7 | A |
| 227 | 41 | 2.6 | 73 | 11.8 | 7.43 | 3.09 | 13.2 | A |
| 228 | 41 | 3.4 | 73 | 12.1 | 7.26 | 3.32 | 18.7 | A |
| 229 | 41 | 4.2 | 73 | 12.3 | 7.74 | 2.94 | 9.6 | A |
| 230 | 41 | 5.4 | 73 | 12.7 | 5.13 | 3.30 | 56.4 | C |
| 231 | 41 | 6.5 | 73 | 13.0 | 6.93 | 3.18 | 19.0 | A |
| 232 | 41 | 4.0 | 73 | 6.0 | -0.897 | 2.600 | 94.4 | E |

The fourth depositional environment is located primarily in the deeper portions of the Central and Western Basins below wave base. Bottom drift is seemingly too weak to actively disturb and rework the sediment deposits. Here the finest grained sandy silts, silts, and clayey silts (Facies C, B, and A) accumulate. Facies B occurs against the western flank of the Mattituck Sill, and thus separates the finer Facies A from the sands of the Eastern Basin. Elsewhere, Facies B and A are probably derived from suspended sediment from the Hudson River entering the Sound via Hell's Gate. Almost one-half of New York City's sewage effluent moves with the surface water into Long Island Sound (M.G. Gross, 1974 personal communication).

The glaciers which helped to shape the Sound also scoured southern New England, leaving little soil to be eroded and car-

Table 6. Summary of Grain-size data by sedimentary facies.

| FACIES | | ≤-2φ | -1φ | 0φ | 1φ | 2φ | 3φ | 4φ | 5φ | 6φ | 7φ | 8φ | 9φ | ≥10φ | AVERAGE φ MEDIAN DIAMETER | AVERAGE φ MEAN DIAMETER | AVERAGE φ SORTING | AVERAGE φ SKEWNESS | AVERAGE PERCENT SAND SIZE |
|---|---|---|---|---|---|---|---|---|---|---|---|---|---|---|---|---|---|---|---|
| | | | | | | WHOLE PHI SIZE VARIABLES | | | | | | | | | | | | | |
| A | 1) | 0.10 | 0.13 | 0.24 | 0.84 | 2.06 | 3.16 | 5.58 | 8.50 | 16.49 | 16.33 | 15.41 | 12.04 | 19.10 | 6.87 | 7.01 | 2.63 | 0.04 | 12.11 |
| | 2) | 0.43 | 0.36 | 0.53 | 1.60 | 2.47 | 2.73 | 3.47 | 4.95 | 5.43 | 5.17 | 3.85 | 2.80 | 9.70 | | | | | |
| B | | 0.00 | 0.08 | 0.16 | 0.25 | 1.63 | 3.85 | 13.43 | 23.55 | 21.60 | 15.17 | 8.74 | 5.26 | 6.27 | 5.33 | 5.71 | 2.02 | 0.19 | 19.40 |
| | | 0.00 | 0.34 | 0.44 | 0.67 | 2.61 | 3.74 | 9.81 | 11.67 | 9.78 | 7.39 | 4.87 | 3.94 | 5.70 | | | | | |
| C | | 0.12 | 0.50 | 0.37 | 1.54 | 9.52 | 20.97 | 13.37 | 9.45 | 13.19 | 9.80 | 7.45 | 5.61 | 8.10 | 4.64 | 4.97 | 2.74 | 0.13 | 46.39 |
| | | 0.46 | 1.74 | 0.46 | 2.63 | 5.66 | 13.24 | 10.12 | 9.55 | 7.71 | 3.86 | 4.13 | 3.37 | 7.21 | | | | | |
| D | | 2.61 | 2.15 | 3.97 | 6.63 | 15.88 | 17.15 | 9.81 | 7.57 | 9.76 | 9.16 | 7.01 | 4.04 | 4.25 | 3.28 | 3.74 | 2.91 | 0.23 | 58.20 |
| | | 4.89 | 3.43 | 4.90 | 4.29 | 9.74 | 11.11 | 9.05 | 6.28 | 5.28 | 5.52 | 3.05 | 3.35 | 4.17 | | | | | |
| E | | 2.49 | 3.09 | 9.71 | 16.61 | 38.85 | 18.08 | 3.13 | 2.44 | 1.51 | 1.53 | 1.06 | 0.78 | 0.72 | 1.38 | 1.55 | 1.37 | 0.11 | 91.96 |
| | | 6.66 | 6.82 | 12.76 | 13.29 | 22.03 | 19.36 | 6.91 | 3.89 | 2.32 | 2.31 | 1.78 | 1.36 | 1.34 | | | | | |

1) Mean
2) Standard Deviation

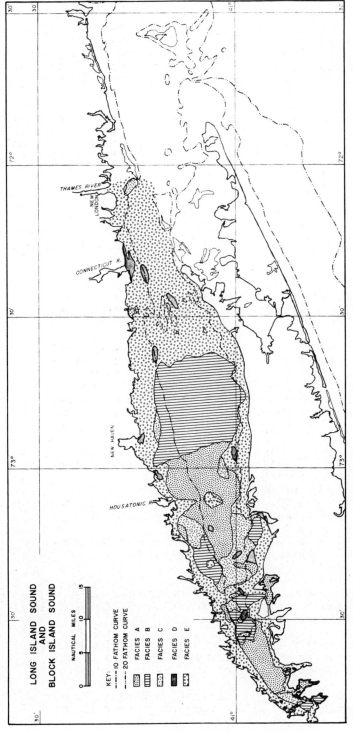

Figure 7. Long Island Sound bottom-sediment facies chart.

ried by rivers. Thus, rivers entering Long Island Sound are not a major source of sediment. None of the rivers have built large deltas from their river-borne materials, nor do extensive beaches occur at their mouths. Seemingly, each estuary acts as a sediment trap, except during flood or storm. Coarse sediment entering the Sound is carried along the shore by tidal currents and longshore drift.

The scarcity of sediment sources feeding Long Island Sound is evidenced by the predoninance of fine-grained deposits (Facies A, B, and C) west of the Mattituck Sill and by the complicated bottom topography which remains visible through the veneer of sediments.

## REFERENCES

Akpati, B.N., 1974, Mineral composition and sediments in eastern Long Island Sound, New York: Maritime Sediments, v. 10, no. 1, p. 19-30.

Ali, S.A., Lindemann, R.H., and Feldhausen, P.H., 1975, Grain size distributions and depositional processes of Great South Bay and South Oyster Bay, New York: Geol. Soc. America, Abstracts with Programs, v. 7, no. 1, p. 21.

Allen, G.P., 1971, Relationships between grain size parameter distribution and current patterns in the Gironde Estuary (France): Jour. Sed. Pet., v. 41, no. 1, p. 74-88.

Buzas, M.A., 1965, The distribution and abundance of Foraminifera in Long Island Sound: Smithsonian Misc. Coll., v. 149, no. 1, 89 p.

Davies, D.K., and Ethridge, F.G., 1975, Sandstone composition and depositional environment: Am. Assoc. Petroleum Geologists Bull., v. 59, no. 2, p. 239-264.

Davis, J.C., 1970, Information contained in sediment-size analysis: Jour. Math. Geology, v. 2, no. 2, p. 105-112.

Dixon, W.J., 1970, Biomedical computer programs: Univ. California Publ. in Automatic Computation, No. 2, Univ. Calif. Press, Berkeley, p. 233-253.

Feldhausen, P.H., 1970, Ordination of sediments from the Cape Hatteras continental margin: Jour. Math. Geology, v. 2, no. 2, p. 113-129.

Feldhausen, P.H., and Ali, S.A., 1974, A multivariate statistical approach to sedimentary environmental analysis: Trans. Gulf-Coast Assoc. Geol. Soc., v. 24, p. 314-320.

Folk, R.L., 1954, The distinction between grain size and mineral composition in sedimentary rock nomenclature: Jour. Geology, v. 62, no. 4, p. 344-359.

Gevirtz, J.L., Park, R.A., and Friedman, G.M., 1971, Paraecology of benthonic Foraminifera and associated micro-organisms of the continental shelf off Long Island, New York: Jour. Paleontology, v. 45, no. 2, p. 153-177.

Goodall, D.W., 1954, Vegetation classification and vegetation continua: Angew. Pflanzensoziologic, Wien. Festschrift Aichinger, v. 1, p. 168-182.

Gordon, R.B., and Pilbeam, C., 1973, Tides and circulation in central Long Island Sound (   .): EOS, Trans. Am. Geophys. Union, v. 54, no. 4, p. 301-302.

Greenwood, B., 1969, Sediment parameters and environmental discrimination: an application of multivariate statistics: Canadian Jour. Earth Sci., v. 6, no. 4, p. 1347-1358.

Grim, M.S., Drake, C.L., Heirtzler, J.R., 1970, Sub-bottom study of Long Island Sound: Geol. Soc. America Bull., v. 81, no. 4, p. 649-666.

Gross, M.G., and Bumpus, D.F., 1972, Residual drift of near-bottom waters in Long Island Sound, 1969: Limnology and Oceanography, v. 17, no. 4, p. 636-638.

Harbaugh, J.W., and Merriam, D.F., 1968, Computer applications in stratigraphic analysis: John Wiley & Sons, Inc., New York, 282 p.

Hardy, C.D., 1972, Movement and quality of Long Island waters, 1971: SUNY Stony Brook, Marine Sciences Research Center, Tech. Report No. 17, 66 p.

Hollman, R., and Sandberg, G.R., 1972, The residual drift in eastern Long Island Sound and Block Island Sound: New York Ocean Science Laboratory (Montauk), Tech. Report No. 0015, 19 p.

Imbrie, J., and Purdy, E.G., 1962, Classification of modern Bahamian carbonate sediments: Am. Assoc. Petroleum Geologists Mem. 1, p. 252-272.

Klovan, J.E., 1966, The use of factor analysis in determining depositional environments from grain-size distributions: Jour. Sed. Pet., v. 36, no. 1, p. 115-125.

Klovan, J.E., and Billings, G.K., 1967, Classification of geological samples by discriminant-function analysis: Bull. Can. Petrol. Geology, v. 15, no. 3, p. 313-330.

Knebel, H.J., and Creager, J.S., 1973, Sedimentary environments of the east-central Bering Sea continental shelf: Marine Geology, v. 15, no. 1, p. 25-47.

Jones, A.S.G., 1971, A textural study of marine sediments in a portion of Cardigan Bay (Wales): Jour. Sed. Pet., v. 41, no. 2, p. 505-516.

McCrone, A.W., Ellis, B.F., and Charmatz, R., 1961, Preliminary observations on Long Island Sound sediments: New York Acad. Sci. Trans., ser II., v. 24, no. 2, p. 119-129.

Merriam, D.F., and Harbaugh, J.W., 1964, Trend-surface analysis of regional and residual components of geologic structure in Kansas: Kansas Geol. Survey Sp. Dist. Publ. 11, 27 p.

Miesch, A.T., and Connor, J.J., 1968, Stepwise regression and nonpolynomial models in trend analysis: Kansas Geol. Survey Computer Contr. 27, 40 p.

Park, R.A., 1968, Paleoecology of *Venericardia sensu lato* (Pelecypoda) in the Atlantic and Gulf coastal province: an application of paleosynecologic methods: Jour. Paleontology, v. 42, no. 4, p. 955-986.

Park, R.A., 1974, A multivariate analytical strategy for classifying paleoenvironments: Jour. Math. Geology, v. 6, no. 4, p. 333-352.

Park, R.A., and Feldhausen, P.H., 1969, Quantitative biofacies analysis: Cape Hatteras North Carolina: Geol. Soc. America, Abstracts with Programs for 1969, pt. 4, p. 60.

Parks, J.M., 1966, Cluster analysis applied to multivariate geologic problems: Jour. Geology, v. 74, no. 5, pt. 2, p. 703-715.

Pautz, M.E., 1969, Severe local storm occurrences, 1955-1967: U. S. Dept. Commerce, ESSA Tech. Memo. WBTM FORST 12, p. 71.

Pierce J.W., and Good, D.I., 1966, FORTRAN II program for standard-size analysis of unconsolidated sediments using an IBM 1620 computer: Kansas Geol. Survey Sp. Dist. Publ. 28, 19 p.

Rao, S.V.L.N., 1968, FORTRAN II program for the calculation of Wilks' $\Lambda$ using an IBM 1620 computer: Kansas Geol. Survey Computer Contr. 20, p. 52-58.

Reyment, R.A., and Ramden, H., 1970, FORTRAN IV program for canonical variates analysis for the CDC 3600 computer: Kansas Geol. Computer Contr. 47, 40 p.

Riley, G.A., 1956, Oceanography of Long Island Sound, 1952-1954, II; Physical Oceanography: Bull. Bingham Oceanogr. Coll., v. 15, no. 1, p. 15-46.

Shepard, F.P., 1954, Nomenclature based on sand-silt-clay ratios: Jour. Sed. Pet., v. 24, p. 152-157.

Sokal, R.R., and Sneath, P.H.A., 1963, Principles of numerical taxonomy: W.H. Freeman and Co., San Francisco, 359 p.

Sorensen, T., 1948, A method of establishing groups of equal amplitude in plant sociology based on similarity of species content and its application to analysis of the vegetation on Danish commons: Biol. Skr., v. 5, no. 4, p. 1-34.

Swift, D.J.P., Sandford, R.B., Dill, C.E., Jr., and Avignone, N.F., 1971, Textural differentiation on the shore face during erosional retreat of an unconsolidated coast, Cape Henry to Cape Hattaras, western North Atlantic: Sedimentology, v. 16, no. 3, p. 221-250.

# SIMULATION TECHNIQUE OF MATCHING AND ITS STABILITY

Dietrich Marsal

*Gewerkschaften Brigitta und Elwerath Betriebsfuhrungs-gesellschaft mbH*

ABSTRACT

The differential equations which govern the performance of natural or technical systems usually contain parameters which are either constants or functions of time and the space coordinates. Unfortunately, these parameters generally are unknown if geological or geophysical processes are involved. However, the parameters may be estimated by a procedure known as matching. The paper discusses an aspect of the validity of matching which shows to be intimately connected with a type of stability. KEY WORDS: *numerical analysis, simulation, geophysics, ground water, hydrology, meteorology, mining, oceanography, petroleum, petrology, sedimentology, soil science.*

INTRODUCTION

Simulation techniques originated in the soil industry. So a brief, descriptive survey of the corresponding problems involved in reservoir engineering is in order. The flow of fluids in the pores and fissures of a reservoir depends mainly on the porosity of the rock, on its permeability to oil, gas, and water, and on the capillary forces acting between the different solid and fluid phases of the producing geological body. If the values of all properties involved are known, then the future performance of oil and gas fields could be predicted rather accurately. Unfortunately, most reservoirs as most other geological bodies constitute an inhomogeneous assemblage of minerals and fluid phases; its properties differ usually considerably from point to point. As a consequence, the variation of qualities and averages are seldom well known. Put into a more mathematical language, this means that we know the equations which govern the flow of oil, gas, and water whereas we do not know with sufficient accuracy the parameters involved.

Nevertheless, the situation is by no means hopeless. Usually, the production of oil, gas, and water is measured regularly at the producing wells, and the reservoir pressure also is known approximately. All these measured values may be interpreted as an actual solution of the corresponding mathematical equations, or, more accurately, as that part of the solution which belongs to the past. The procedure which tries to estimate the parameters of the equations from a known part of the solution is known as "history matching".

## HISTORY MATCHING

Actually, history matching is performed either by using special formulae adopted to the peculiar mathematical equations involved or by trial and error. In this instance, the matching is done by choosing first a set of parameters which seems to be reasonable with respect to all measurements and observations. Then, this set of parameters is incorporated into the equations, and a computer run is made which calculates the production of the past. As the last step of this first matching cycle the computed data are compared with the actual history. If the difference is significant, then a second set of parameters is chosen and the procedure is repeated. This goes on until a sufficient matching is accomplished or until a further improvement does not seem to be within reach. Now and then this procedure of trial and error is automized by using the method of least squares or a random-number generator.

A similar situation may arise whenever an inhomogeneous body is studied quantitatively. As examples, movements of subsurface water, diagenetic changes of all types in sediments, physical and chemical processes in water bodies as oceans and lakes, and the study of the distribution of temperature and pressure in parts of the earth's crust and in planets and stars may be cited. Here again the matching may be a matching of the past as a first step of a forecasting, or the matching may refer to a steady-state process, estimating the properties of an inhomogeneous body from well-known values measured at accessible locations.

Actual matching performed in reservoir engineering is somewhat laborious because a rather large number of parameters and some parameter functions have to be estimated. On the other hand, the matching is done in a naive manner. There are no investigations whether the results obtained by matching are unique. It may be that an infinity of different parameter values exist, each set of values yielding more or less the correct history of the past. Consequently, a successful matching does not guarantee that the inhomogeneities of the mathematical model match the inhomogeneities of the geological body. This means with respect to reservoir engineering that a correct matching of the past may result in a wrong forecast of future production. However, the reliability of a forecast will improve as the time span of the matched past increases.

## AN EXAMPLE

A numerical example of matching and its reliability, is given by a heat-generating radioactive substance with large half-life which is distributed irregularly in a rock with constant thermal conductivity. It is a two-dimensional, steady-state situation assuming a state of balance between the generation of heat within the geological body and the loss of heat to the environment. If the thermal conductivity and the heat production in the source function $S(x,y)$ are combined, then temperature $T$ satisfies the partial differential equation

$$T_{xx} + T_{yy} + S(x,y) = 0.$$

For simplicity let the solid be a square with unit length of edge and known boundary temperatures. A square grid with 36 grid points, containing 16 interior grid points is used (Fig. 1).

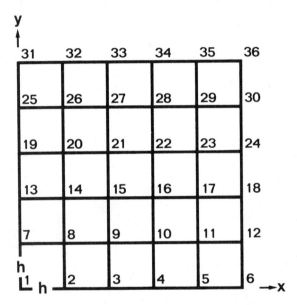

Figure 1. Grid with 36 points and 16 interior points.

Then we replace the derivatives by divided differences, using the usual approach so that the temperature at a grid point is put equal to the mean of the temperatures of its nearest neighbors plus a term taking into account the heat production. At each interior grid point the temperature is unknown, and each interior grid point gives rise to one equation. Therefore the result is 16 equations containing 16 unknowns. Two typical equations are

grid point 15: $\quad -4T_{15} + T_9 + T_{16} + T_{21} + T_{14} = -h^2 S_{15}$

grid point 23: $\quad -4T_{23} + T_{17} + T_{29} + T_{22} \quad\quad = -h^2 S_{23} - T_{24}.$

If the origin of the coordinate system is located at the grid point 1 of Figure 1, and if $S = x^2 + xy + y^2 + x + y$, then the system of difference equations provides the following distribution of temperature, rounded to three significant figures. The omitted values at the corners of the square are not needed to calculate the temperatures at the interior points:

```
     4*    5*    6*    7*
3*   3.56  4.39  5.29  6.18  7*
2*   2.78  3.62  4.48  5.29  6*
1*   1.88  2.76  3.62  4.39  5*          *known boundary values
0*   0.95  1.88  2.78  3.56  4*
     0*    1*    2*    3*
```

Now assume that the local variations of the thermal source function S are not known. However, the temperature measured at the grid point 15 is $T_{15} = 2.7637$, and from this poor information the values of T and S at the other interior grid points are to be determined. The system of difference equations have to be solved by using a set of numbers, $S_8, S_9, \ldots, S_{29}$, that $T_{15}$ matches more or less accurately the value of 2.7637.

Is this strategy reasonable? If $T_{15}$ is replaced by the known value 2.7637, a system of 16 equations containing 15 unknown values of temperature and 16 unknown values of the source function is obtained. Therefore there are an infinite number of S-Values which all comply with the requirement that $T_{15} = 2.7637$. Nevertheless a matching is not meaningless if the grid parameter $h^2$ is smaller than 1. In this situation a considerable change of say $S_{15}$ does not change $h^2 S_{15}$ much. Consequently, if the system of difference equations is not sensitive to rounding errors, a distinct change of $S_{15}$ does not have much effect on the solution of the system. The estimated temperatures are rather insensitive to errors of the heat-source function. On the other hand, in this situation a good matching of temperatures does not necessarily imply a good matching of the heat-source function. Therefore when matching care should be taken in drawing conclusions which necessitate a more or less accurate knowledge of the inhomogeneities of a region under consideration. To estimate the possible errors involved, a numerical computer experiment should be performed choosing different model inhomogeneities and studying the effect of these alterations on the solution of the equations.

## RESULT

In a numerical example, the test function used was

$$S = Ax^2 + Bxy + Cy^2 + Dx + Ey$$

$0 \leq A \leq 2$      $0 \leq B \leq 2$      $0 \leq C \leq 2$

$0 \leq D \leq 2$      $0 \leq E \leq 2$

with undetermined coefficients A, B, C, D, and E. A random-number generator chose the coefficients, and a program calculated the corresponding S-values and corresponding solution of the difference quations. We performed 100 runs. The biggest absolute deviation of the T-values at all grid points was less than 0.1, the biggest relative deviation was 3 percent. On the other hand, the generated S-values differed as much as 60 percent from the correct values. A poorly approximated source function resulted in a good match of temperature at every grid point.

# MATHEMATICAL MODELING OF SEDIMENT ACCUMULATION IN PROGRADING DELTAIC SYSTEMS

Daniel H. Horowitz

*Exxon Production Research Company*

ABSTRACT

Thickness and sedimentation rates of time-stratigraphic units deposited during deltaic progradation can be predicted from a mathematical model. Depositional surfaces are constrained to lie along an assumed equilibrium bathymetric profile that advances seaward at a specified rate. Subsidence is treated as an isostatic response and so is related to the weight of accumulated sediment. Effects of contemporaneous compaction are included to obtain correct weights and thicknesses of stratigraphic units.

For a uniformly prograding deltaic system with characteristics similar to that in the northern Gulf of Mexico, the model predicts that sedimentation rate at a fixed locality will continually increase as long as delta slope sediments are deposited. This is one of the reasons abnormally high fluid pressures and faults occur in offshore Louisiana. KEY WORDS: *simulation, abnormal fluid pressure, sedimentology, stratigraphy.*

INTRODUCTION

Geologists who study the sedimentary filling of basins are familiar with the classical sketches depicting prograded deltaic deposits as a series of seaward imbricated wedges (Fig. 1). By quantifying the mechanical response of the basin to sedimentary loading and mathematically describing the equilibrium profile at the leading edge of the deltaic mass, it should be possible to advance from this simple graphical illustration to a mathematical model that more realistically describes sediment accumulation during deltaic progradation. The results of the model then could be used to provide further insight into such problems as the evolution of fluid pressure during deltaic advance, zonation of slope deposits into time-stratigraphic units, and motivation for geosynclinal subsidence.

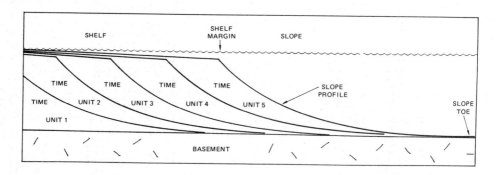

Figure 1. Classical representation of time-stratigraphic units in prograded deltaic complex as series of seaward imbricated wedges. Delta profile is divided into shelf and slope segments that join at shelf margin.

Several such models already have been devised by Harbaugh and Bonham-Carter (1970). Many simplifications understandably were made in their pioneering effort. This report seeks to place their models on a more realistic footing by including the effects of contemporaneous compaction and specifying subsidence rigorously as an isostatic response instead of arbitrarily. The concept of an equilibrium delta profile also is newly incorporated so that measurable geologic information can be introduced into the models.

## BASIN RESPONSE TO LOADING

Basin response to loading is represented by a simplified isostatic model illustrated in Figure 2. The left-hand columnar section in the figure depicts a predeltaic situation and is idealized to show a deep-water column directly overlying continental crust or some other relatively uncompactable sediment. The right-hand columnar section depicts a subsequent situation when deltaic slope deposits of thickness H have accumulated. The thickness H is the sum of two components, F which represents basin fill or shallowing, and S which represents isostatic subsidence:

$$H = S + F \qquad (1)$$

Assuming isostatic equilibrium prevails, the weight of both columns should be equal at the compenation level. Using the symbols defined in Figure 2,

$$\rho_W \cdot WD_o + \rho_C \cdot (\text{Crustal Thickness}) + \rho_M S$$
$$= \rho_W WD + \rho_C \cdot (\text{Crustal Thickness}) + (\text{Weight Thickness H of Sediment})$$

which can be simplified to

$$\rho_W(WD_o - WD) + \rho_M S = \text{Weight Thickness of Sediment} \qquad (2)$$

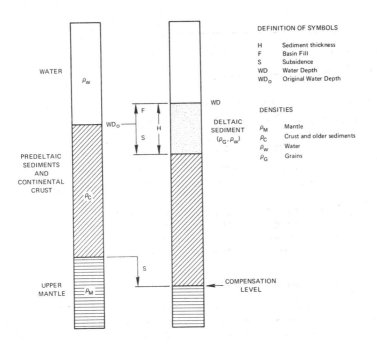

Figure 2. Simplified isostatic model employed in calculations, and list of symbol definitions. Left column showing predeltaic situation and right column showing situation during deltaic deposition should weigh same at compensation level if both columns are in isostatic equilibrium.

To obtain the weight of sediment requires a knowledge of the bulk density or porosity profile through the sediment, because sediments usually become less porous with depth.

Because shaly sediments typical of those that might be deposited on deltaic slopes generally exhibit a progressive and uniform porosity loss with depth (Baldwin, 1971), it is not unlikely that a simple function can be determined that will approximate the relationship between porosity ($\phi$) and subsurface depth (Z) for an area of interest. Let $\phi(Z)$ be that function. The volume of solid material and voids in a small thickness dZ of sediment at depth Z then is given by $[1-\phi(Z)]dZ$ and $\phi(Z)dZ$, respectively. Multiplying these infinitesimal volumes respectively by grain and fluid densities $\rho_G$ and $\rho_W$ (assume the voids are filled with water), and integrating over the depth range 0 to H, yields the bulk weight of sediment:

$$\text{Weight Thickness H of Sediment} = \rho_W \int_0^H \phi(Z)dZ + \rho_G \int_0^H [1-\phi(Z)]dZ \quad (3)$$

The relationship characterizing basin response to loading is obtained by substituting (3) into (2):

$$\rho_W(WD_0-WD) + \rho_M S = \rho_W \int_0^H \phi(Z)dZ + \rho_G \int_0^H [1-\phi(Z)]dZ \qquad (4)$$

The S term may be eliminated with the help of (1) and the further substitution of $(WD_0-WD)$ for F, because the sediment fill, F, merely represents the amount of basin shallowing:

$$S = H - (WD_0-WD) \qquad (5)$$

With this substitution and further simplification, equation (4) becomes

$$\int_0^H \phi(z)dz = \frac{\rho_M-\rho_W}{\rho_G-\rho_W}(WD_0-WD) - \frac{\rho_M-\rho_G}{\rho_G-\rho_W} \cdot H \qquad (6)$$

The equation relates thickness H of sediment to water depth WD, given initial water depth $WD_0$, densities of components involved, and the porosity-depth function $\phi(Z)$, and assuming isostatic equilibrium. In this study H is the dependent variable relative to WD, whose values can be estimated from other geological information. A trial and error method must generally be used to solve for H, but with a computer an iteration scheme can be employed.

## DELTA PROFILE GEOMETRY

Water depth, the independent variable, is specified as a locus of points on a bathymetric profile along the slope or leading edge of a deltaic wedge. This slope profile is typically drawn as a concave-upward surface more steeply inclined at the shelf margin and asymptotically approaching a maximum water depth seaward (Fig. 1). To quantify this shape, a coordinate system with depth, WD, on the vertical axis (positive downward) and seaward distance, x, on the horizontal axis, is adopted (Fig. 3). The slope profile then is represented by a function $WD(x,t)$, where t is time. Time is an independent parameter that governs changes in position and shape of the slope profile during deltaic advance. $WD(x,t)$ is assumed to represent an equilibrium bathymetric profile that is maintained (within the limits specified) during contemporaneous sedimentation, subsidence, and compaction.

In its most general form the equation representing the curve $WD(x,t)$ contains four terms which define terminal points on the slope profile. The landward termination occurring at the shelf margin has coordinates $M(t)$, $WD_m(t)$, where $M(t)$ is the seaward position of the shelf margin and $WD_m(t)$ is its water depth. The seaward termination at the slope toe is similarly defined by coordinates $T(t)$ and $WD_t(t)$ as shown in Figure 3. The $M(t)$ term governs rate of deltaic progradation, and $T(t)$,

# MATHEMATICAL MODELING OF SEDIMENT ACCUMULATION 109

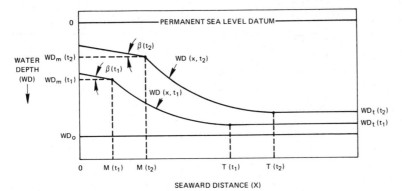

Figure 3. Coordinate system and parameters employed to describe delta profile. Profiles at two different times are illustrated. See text for definition of parameters.

which indicates rate of toe advance, equals M(t) unless the profile length changes with time. $WD_t(t)$ describes the amount of deep water aggradation due to pelagic deposition, and $WD_m(t)$ indicates the amount of shelf aggradation.

Specifying a slope profile and its rate of advance marks a departure from the approach used by Harbaugh and Bonham-Carter (1970). In their simulations the amount of sediment added and its fractionation seaward were stipulated. This information ordinarily is not known and limits the practical applicability of their models. Slope profiles, on the other hand, can be constructed from bathymetric profiles (admittedly this applies only for present not past profiles), and rate of deltaic advance can be reconstructed from geologic information, e.g., see facies and environmental maps published by Lafayette and New Orleans Geological Societies (1968).

Equations for two elementary concave upward curves that might qualify as slope profiles, an exponential curve and an arc of a circle, are listed:

SLOPE PROFILES

EXPONENTIAL:

$$WD(x,t) = [WD_t(t) - WD_m(t)]e^{-k[x-M(t)]} - WD_t(t) \quad x \pm M(t) \qquad (7)$$

CIRCULAR:

$$WD(x,t) = WD_t(t) - k(t) + \sqrt{k(t)^2 - [T(t)-x]^2} \quad M(t) \geq x \geq T(t) \qquad (8)$$

$$k(t) = \frac{[T(t)-M(t)]^2 + [WD_t(t)-WD_m(t)]^2}{2[WD_t(t)-WD_m(t)]}$$

The term k which controls the exact shape of the curve is somewhat arbitrary for an exponential curve but functionally dependent on other terms for a circular arc. The exponential curve theoretically extends to infinity and only approaches $WD_t$ in the limit, whereas the circular arc has a finite seaward termination. Both are valid only for $x \geq M$ because a different equilibrium profile applies for shelf sediments. To a first approximation, the shelf profile can be described by a line uniformly sloping seaward at angle $\beta$:

SHELF LINE PROFILE: $WD(x,t) = WD + [x-M(t)]\tan\beta \quad x<M(t)$ (9)

Not explicitly considered is the effect of sea-level fluctuations. Strictly speaking, therefore, all water depths are referred to permanent sea-level datum or some other fixed datum, such as average sea level.

## PREDICTING THICKNESSES OF TIME-STRATIGRAPHIC UNITS

The rate at which the sediment column at a certain locality (x fixed) grows can be predicted by inserting the slope profile equation $WD(x,t)$ into equation (6), allowing t to change through a specified range, and computing H, now also a function of x and t, at different times. It is necessary, of course, first to stipulate how $M$, $WD_t$, $WD_M$, and T differ with time so that $WD(x,t)$ is defined. Once sediment thicknesses $H(x,t)$ for different time increments are obtained, the amount of subsidence S is calculated from equation (5), and the amount of basin fill F is simply the difference between original and calculated water depths,

$$F = WD_0 - WD(x,t). \quad (10)$$

Predicting the thickness of time stratigraphic units in a sedimentary column is not as easy. This is because each unit in the column compacts at a different rate during burial, shallow units compacting more rapidly than deeper ones in accordance with the porosity-depth function $\phi(Z)$. The previously computed variable H expresses the thickness of the total sediment column and does not directly reveal unit thicknesses. It therefore is necessary to define a new variable which depends on sediment accumulated in individual units rather than in a composite sediment column.

That new variable is the net or water-free thickness of a unit, $NT_j$, where the subscript j identifies the unit. For sediments it is a conceptual measure that represents the thickness a unit would have if all its voids were eliminated by compaction and bulk density equaled grain density. Although NT is not the stratigraphic thickness of a time-stratigraphic unit, it will yield that value once the proper amount of water-filled void space is added.

Unit net thickness $NT_j$ is derived from the sediment-column thickness H as sketched in Figure 4. Let,

$$H(x,t_1), H(x,t_2) \ldots H(n,t_n)$$

represent a series of sediment thicknesses at locality x and times $t_1, t_2, \ldots t_n$. For each H value there corresponds a net ickness

# MATHEMATICAL MODELING OF SEDIMENT ACCUMULATION

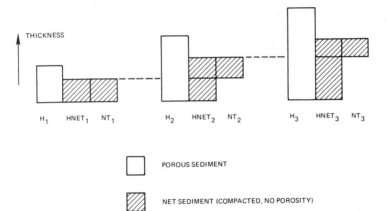

Figure 4. Relationship between successive H, HNET, and NT variables needed to calculate thicknesses of time-stratigraphic units.

HNET$(x,t_1)$, HNET$(x,t_2)$ etc., obtained by integrating the volume of solids over the thickness interval,

$$HNET(x,t_j) = \int_0^{H(x,t_j)} [1 - \phi(Z)]dZ \qquad j = 1, 2, \ldots, n \qquad (11)$$

Successive values of HNET, like those of H, incorporate the sediment contribution of successively younger time stratigraphic units, but the manner of incorporation is so basically different that $NT_j$ values may be calculated directly from the series of HNET values but not from those of H. Whereas each H value depends on the gross compactional-depositional response of a composite system, successive HNET values differ by the net thickness or solids contribution of successive units, and because net thickness is invariant with time or burial depth, it is unique to a unit. Accordingly, net thicknesses of successive units are given by

$$NT_j(x) = HNET(x,t_j) - HNET(x,t_{j-1}) \qquad j = 1, 2, \ldots, n \qquad (12)$$
$$HNET(x, t_0) = 0$$

Before indicating the manner in which $NT_j$ is converted to stratigraphic thickness, the significance of $NT_j$ values themselves is noted. They essentially represent the net rate of sedimentation, $RS_j$, in the time period involved:

$$RS_j(x) = \frac{NT_j(x)}{t_j - t_{j-1}} \qquad (13)$$

If $RS_j$ is multiplied by grain density $\rho_g$, the net rate of sedimentation in mass per unit time is obtained. This manner of calculating sedimentation rates is recommended instead of the more common method of dividing a unit's thickness by its age span, which suffers by being functionally dependent on the unit's porosity.

Determining thickness or subsurface depths of time-stratigraphic units in a sedimentary column at time $t_j$ requires a procedure that is reverse of calculating net thickness. Starting with the known net thickness $NT_j$ of the youngest unit, conceptually "pump in the water" stipulated by the porosity-depth relationship $\phi(Z)$ and incrementally extend the sediment column downsection by repeating the procedure for successively older units. The procedure is accomplished as follows.

Set up the equation which expresses the relationship at locality x and at time $t_j$ existing among the net thickness $NT_j$ of the youngest unit j, the porosity-depth function $\phi(Z)$, and the top ($DEPTH_j=0$) and basal ($DEPTH_{j-1}$) depths of the unit:

$$NT_j(x) = \int_{DEPTH_j(x)=0}^{DEPTH_{j-1}(x)} [1-\phi(Z)]dZ \qquad (14)$$

The base of the unit is at $DEPTH_{j-1}$, which is the depth to which a net-thickness unit would "expand" if the porosity distribution stipulated by $\phi(Z)$ existed. $DEPTH_{j-1}$ is the only unknown in the equation so its value is determinate. A trial and error approach or iteration scheme, however, may be required to determine its value. With the base $DEPTH_{j-1}$ obtained, the next oldest unit j-1 is considered:

$$NT_{j-1}(x) \int_{DEPTH_{j-1}(x)}^{DEPTH_{j-2}(x)} [1-(Z)]dZ$$

Here too, only the upper limit of integration is unknown. The procedure is repeated until the basal depth of the oldest unit is determined. The succession of depths mark the tops of progressively older time stratigraphic units existing at the specified time $t_j$ and locality x.

## EXAMPLE

To illustrate the procedure and provide a basis for discussing some of the applications of the technique, an example is presented. Input variables and functions were selected to correspond to a prograding deltaic system comparable in size and character to that in the Louisiana offshore area of the Gulf of Mexico. A circular slope profile was assumed and the following values were used:

$\rho_w = 1.0$ gm/cc

$\rho_M = 3.3$ gm/cc

$\rho_G = 2.7$ gm/cc

$WD_0 = 3000$ m

$WD_M = 150$ m

$T(t) - M(t) = 150$ km         $t$ = time in years

$M(t) = .006t$ m

$\beta = 0°05'$

$\phi(Z) = 0.7/(1+.001Z)$         $Z$ = depth in meters

With these choices the delta profile function WD(x) is:

$$WD(x,t) = 3000 - 3.95 \cdot 10^6 + \sqrt{15.6 \cdot 10^{12} - [1.5 \cdot 10^5 + .006t - x]^2} \quad .006 \leq x \leq 1.5 \cdot 10^5 + .006t$$

$$= 3000 \quad\quad\quad\quad\quad\quad\quad\quad\quad\quad\quad\quad x > 1.5 \cdot 10^5 + .006t \quad\quad (15)$$

$$= 150 - 1.455 \cdot 10^{-3}(.006t - x) \quad\quad\quad\quad x < .006t$$

Some of the choices deserve comment. The water depth at the slope toe and beyond is maintained at 3000 m, which means pelagic sedimentation is considered negligible. Rate of deltaic advance M(t) is maintained at 6 mm per year, and whereas this is slower than modern deltaic lobes prograde it is typical of the collective rate of advance exhibited by coalescing deltaic systems (Curtis, 1970). Finally, the porosity function selected closely approximates the porosity-depth relationship obtained by Bryant, Hoffman, and Trabant (1974) when they compacted shaly sediments from the floor of the modern Gulf of Mexico in the laboratory (Figure 5).

To perform the calculations the delta-profile function WD(x,t) in (15) is substituted for WD in equation (6). A locality is selected by fixing x, and the time variable t is allowed to range through a desired time interval. By changing x also, results at different locations are obtained (if basement geology changes laterally, a different isostatic equilibrium relationship may apply for each locality). Solving (6) yields a series of values H(x,t) from which all other measures, ultimately the depths of time-stratigraphic units, are derived using in succession equations (11), (12), and (14). Results may be viewed from two perspectives depending on whether x is fixed or permitted to change.

By allowing x to take on a range of values, the geometry of time-stratigraphic units in a deltaic wedge is revealed. The cross-section appearance of time-unit boundaries differing in age by five million years is sketched in Figure 6. In this example it was assumed that crustal thickness and density are uniform throughout the length of the cross section, so the same isostatic equilibrium relationship (6) applies for all localities. The patterns revealed by Figures 1 and 6 are generally similar aside from subsidence effects also included in Figure 6. However,

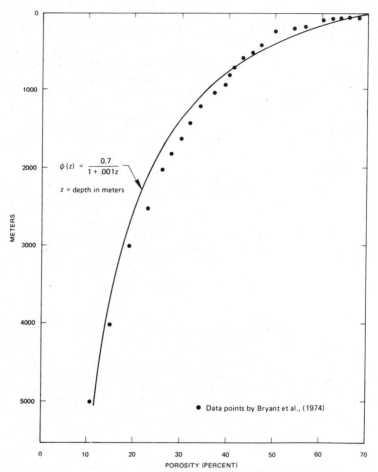

Figure 5. Porosity-depth function used in example closely approximates data points obtained by Bryant, Hoffman, and Trabant (1974) based on laboratory compaction of shaly Gulf of Mexico sediments.

patterns drawn in Figure 1 are rather arbitrary, whereas those in Figure 6 are defined mathematically and potentially realistic provided correct values are assigned to variables and assumptions are valid. The abrupt and unrealistic change in geometry at the 300-km mark, incidentally, results because the following two simplifying but probably invalid assumptions were made: (1) isostatic subsidence acts in response to a vertical-line load instead of an average load distributed over a broad area; and (2) the shelf margin is defined by the intersection of two profiles meeting at an acute angle instead of a smooth transition profile.

By holding x constant the history of sediment accumulation at a given locality is revealed. Figure 7 is a time scan which shows

# MATHEMATICAL MODELING OF SEDIMENT ACCUMULATION    115

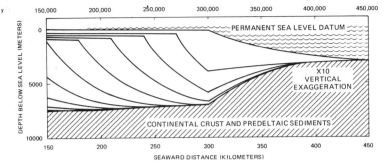

Figure 6. Geometry of time-stratigraphic units predicted from mathematical model assuming uniform rate of progradation and homogeneous crust and predeltaic sediment. Time-line unit boundaries are spaced five million years apart. Time zero corresponds to time when shelf margin and slope toe were located at seaward distances x=0 and x=150 km, respectively. Section between these distances is omitted because pre-deltaic sediments and transient delta profiles would have to be specified.

the history of subsidence and fill, relative to an assumed permanent sea-level datum, at the locality along the left-hand margin of Figure 6. Also shown in Figure 7 are the compaction-corrected, time-depth trajectories of time-stratigraphic units. The fact that predicted rates of sedimentation increase as the shelf margin approaches can be inferred from Figure 7, but in Figure 8 these rates are plotted on a semilogarithmic scale using equation (13).

## INFERENCES BASED ON THE MODEL

Before discussing some of the implications and applications of this modeling technique, factors that might be important but which were omitted in this first attempt should be noted. These include basement inhomogeneities, contemporaneous structuring, sea-floor topography, time effects on porosity and isostatic adjustment, sea-level fluctuations, mantle phase changes, and no doubt others. Thus, only some of the gross aspects of deltaic sedimentation can be revealed by this modeling technique, and all inferences that are drawn must be tentatively regarded.

One result of the model calculations, the predicted rates of sedimentation shown in Figure 8, is interesting in that it helps explain the origin of abnormally high-fluid pressures in areas with prograding deltaic complexes, such as offshore Louisiana and the Niger Delta. According to Rubey and Hubbert (1959), rapid loading of relatively impermeable sediment is likely to lead to overpressuring. Slope deposits are relatively shaly and therefore

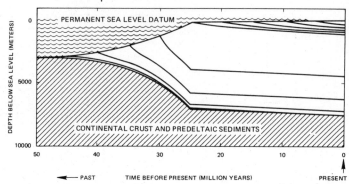

Figure 7. History of sediment accumulation, subsidence, and bathymetry at fixed locality (locality x=150 km in Figure 6) is revealed by time-depth trajectories of time-line unit boundaries. Present stratigraphy is plotted on right-hand margin, and progressively older bathymetric-stratigraphic columns are shown to left. Note that each time-stratigraphic unit becomes thinner as it is buried (to right) owing to compaction.

should have low permeabilities after moderate compaction. From Figure 8 it is evident that the other requirement, rapid loading, also exists during slope deposition, for rate of deposition during uniform deltaic advance increases over two orders of magnitude in passing from delta-toe to shelf-margin deposition. Unless some key assumptions need to be reconsidered, the relative change in predicted sedimentation rate may prove realistic for some natural delta systems. In offshore Louisiana where rate of deltaic advance has increased with time, the disparity in sedimentation rates may be higher yet.

An additional inference regarding the evolution of overpressuring can be drawn if the relative increase in sedimentation rate is realistic. It seems likely that the slowly accumulating deltaic toe deposits would start with initially normal pressure and only later become overpressured when sedimentation rates become excessive. Consequently these sediments may be compacted even though currently existing pressures are sufficient to support most or all of the overburden pressure that causes compaction. Upper slope deposits, on the other hand, should become overpressured sooner because they are deposited while sedimentation rates are highest; hence, they may be less compacted, after adjustment for depth, than toe deposits.

If amount of overpressuring in a section generally increases in conformity with the increase in sedimentation rate, peak pressures should occur during uppermost slope deposition (Fig. 8).

# MATHEMATICAL MODELING OF SEDIMENT ACCUMULATION

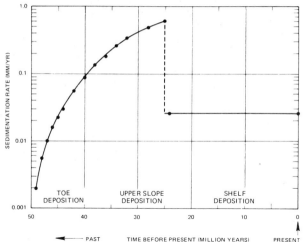

Figure 8. Sedimentation rate versus time, plotted on semilogarithmic scale (see Figure 7). Sedimentation rate represents net thickness of sediment (i.e., totally compacted and water free) deposited per unit time. Multiplying plotted sedimentation rate by grain density $\rho_G$ would yield mass rate per unit time.

Because high-fluid pressures assist in the initiation and movement of faults (Handin and others, 1963), the slope margin should be rather unstable tectonically. Bruce (1973) calls attention to a relationship between development of high fluid pressures and the increased susceptibility of sediments to faulting.

Another stimulating result of the model study is the predicted thickness of shelf deposits if sediments have aggraded to sea level (left-hand part of Figure 6). Although dimensions and attributes in the mathematical model were selected to simulate an offshore Louisiana deltaic system, predicted shelf-deposit thicknesses of about a thousand meters are only a third or fourth those in that area (see figure 4 in paper by Lafayette and New Orleans Geological Societies, 1968). A number of reasons can be offered for the discrepancy, starting with the admission that the model with all its simplifications is inadequate. Aside from that, unaccounted volume losses in preshelf rocks should be considered. These may include dewatering of abnormally pressured shales along faults, withdrawal and solution of salt, and phase transitions which increase the density of material in the upper mantle. A time lag in isostatic response could explain late-stage subsidence also, but Crittenden's (1963) study of isostatic rebound in Pleistocene Lake Bonneville indicates the response is geologically rapid - several thousand years compared to the millions during which shelf sediments in offshore Louisiana have accumulated. Use of one porosity-depth curve for normally pressured shale, abnormally pressured shale, and interbedded sand and shale in the shelf deposits also introduces some error but not near enough to explain the discrepancy.

## GENERAL APPLICATIONS

For applications in other areas and for more detailed studies in offshore Louisiana, geologic and bathymetric data should be employed to assign values for the input parameters. Basement geology can be inferred from geophysical data, and former positions of shelf margins can be determined from paleobathymetric maps based on shallow well control. These maps may be consulted also to determine shelf margin water depths and to help reconstruct the upper part of the slope profile, but in many problems current slope profiles based on bathymetric maps will have to be used and the assumption made that past and present profiles are similar. Porosity-depth information can be assembled from log data or inferred from changes in seismic velocity with depth.

Implicit in any application is the assumption that the mathematical model realistically describes sediment accumulation in a prograding deltaic system. Considering the model's simplicity it probably only roughly approximates the real world. That, however, is not the important point. What is significant is that a formally structured sedimentological model explicitly specifying controlling factors and describing their interaction has been devised. As such the model is capable of improvement because its parts can be examined and then modified or expanded as our understanding of earth processes advances.

## REFERENCES

Baldwin, B., 1971, Ways of deciphering compacted sediments: Jour. Sed. Pet., v. 41, no. 1, p. 293-301.

Bruce, C.H., 1973, Pressured shale and related sediment deformation: mechanism for development of regional contemporaneous faults: Am. Assoc. Petroleum Geologists Bull., v. 57, no. 5, p. 878-886.

Bryant, W.R., Hoffman, W., and Trabant, P., 1974, Permeability of unconsolidated and consolidated marine sediments, Gulf of Mexico, *in* Study of naturally occurring hydrocarbons in the Gulf of Mexico: Texas A&M Univ., College Geosciences, 108 p.

Crittenden, M.D., Jr., 1963, New data on the isostatic deformation of Lake Bonneville: U.S. Geol. Survey Prof. Paper 454-E, 31 p.

Curtis, D.M., 1970, Miocene deltaic sedimentation, Louisiana Gulf Coast, *in* Deltaic sedimentation, modern and ancient: Soc. Econ. Paleontologists and Mineralogists Sp. Publ. 15, p. 293-308.

Handin, J., Hager, R.V., Jr., Friedman, M., and Feather, J.N., 1963, Experimental deformation of ordinary rocks under confining pressure: pore pressure tests: Am. Assoc. Petroleum Geologists Bull., v. 47, no. 5, p. 717-755.

Harbaugh, J.W., and Bonham-Carter, G., 1970, Computer simulation in geology: John Wiley & Sons, Inc., New York, 575 p.

Lafayette and New Orleans Geological Societies, 1968, Geology of natural gas in South Louisiana, *in* Natural gases in North America, v. 1: Am. Assoc. Petroleum Geologists, Tulsa, Oklahoma, p. 376-581.

Rubey, W.W., and Hubbert, M.K., 1959, Role of fluid pressure in mechanics of overthrust faulting: Geol. Soc. America Bull., v. 70, no. 2, p. 167-206.

# A SEDIMENTOLOGICAL PATTERN RECOGNITION PROBLEM

Malcolm W. Clark and Isobel Clark

*University of London*

ABSTRACT

Analysis of grain-size distributions of coastal sands reveals that the distributions may be considered as composed of two (or more) lognormal components. It is tempting to infer that these components are derived from different depositional mechanisms. More faith could be placed in this inference if other characteristics of the deposit were describable in terms of the mixing of a similar number of components. Attention has been directed to the shape characteristics of the deposits. Feature extraction was achieved by digitizing the perimeter of the silhouettes of about 700 grains and fitting a truncated Fourier series to the outlines. The first eight harmonic amplitudes of these series were analyzed to detect naturally occurring clusters. Nonlinear mapping, fuzzy-set analysis and multivariate-mixing analysis were employed to determine clusters. KEY WORDS: *data display, mapping, cluster analysis, discriminant analysis, Fourier analysis, fuzzy-set analysis, multivariate mixing, statistics, sedimentology.*

INTRODUCTION

It has long been realized by sedimentologists that there is a correspondence between the size-frequency distributions of many sediments and the lognormal distribution. Some arguments have been advanced, notably by Middleton (1970) and Mahmood (1973), to provide some theoretical justification for this observation. However, despite these arguments, observed distributions continue to be wayward, and the correspondence between observed and theoretical distributions is convincing in only a few of the reported situations. In order to account for these discrepancies, suggestions have been made to employ probability densities other than the lognormal. Tanner (1958) examined the Pearson Type I and IV distributions, Krumbein and Jones (1970) used a Gamma distribution, and Bagnold (1941) suggested the use of a function akin to the lognormal; Kittleman (1964) applied the Rosin-Rammler distribution.

Although these alternative distributions seem to provide a closer fit, they lack the general applicability of the lognormal, and, with the exception of the Rosin-Rammler, seem to have no theoretical justification. The Rosin-Rammler can be derived for crushed materials, and thus should apply to broken, unsorted rock material.

An alternative, which retains the generality of the lognormal, but introduces more flexibility, is to regard the frequency distribution as a sum of several lognormals, or, because the problem may be specified in terms of logarithms, a sum of normals

$$q(x;\underline{\theta}) = \sum_{i=1}^{i=m} \alpha_i \phi\left[\frac{x-\mu_i}{\sigma_i}\right] \quad (1)$$

where m is the total number of components;
$\mu_i$ is the mean of the $i$th component distribution;
$\sigma_i$ is the standard deviation of the $i$th component distribution;
$\alpha_i$ is the proportion of the overall popultation deriving from component distribution $i$;
$\phi(z)$ is the probability density function of the standard normal distribution; and
$\underline{\theta}$ is a vector of parameters $(\mu_1 \sigma_1 \alpha_1 \mu_2 \sigma_2 \alpha_2 \ldots \mu_m \sigma_m \alpha_m)$.

This model has been discussed in sedimentological terms by Tanner (1964), Spencer (1963), and Folk (1971), among others. A similar model has been employed by Visher (1969), where he considers a sediment-size frequency curve to be composed of a sequence of truncated lognormal components. A similar model has been proposed by McKinney and Friedman (1970), where they extract a major lognormal component, and two smaller subsidiary components which need not be lognormal. Doeglas (1946) also suggested an additive scheme, but maintained that the underlying components were normal, not lognormal. This conflict can be resolved to some extent, because one of his reasons for choosing normal components was that coarse deposits commonly have symmetric size-frequency distributions. It can be demonstrated that as the mean of a lognormal distribution increases, its skewness decreases, until it becomes an essentially symmetric distribution.

There are some problems associated in this type of size-component analysis, the major problem lies in determining how many components are needed to approximate the observed distribution adequately. Whereas Walger (1961) suggested that no deposit is composed of more than three lognormal components, both van Andel (1973) and Curray (1960) published accounts where more than three components are present. One reason for this contradiction is that no obvious method exists of deriving sample size from weight-frequency data, so that "goodness of fit" tests, such as $\chi^2$, can be applied. Jones (1969) discussed this problem, and suggested that a value for the sample size may be determined considering the "smallest reproducible weight" of the weighing procedure, and using this as the basis for the total number of units in the frequency distribution. Using this convention it is possible to fit a model, using some objective criterion to decide if sufficient components have been fitted. The multitude of components observed by van Andel and Curray also may be attributed to the fact that they were taking offshore samples, where control to sample only a single sedimentation unit would have been impossible; thus their deposits were likely to be mixtures of several layers.

Fitting the model also may be a problem, but Clark (1976) suggested that a variety of numerical techniques may be employed. In analyzing about 120 beach and dune samples, it was determined that a two-component lognormal model provides a satisfactory fit in almost all situations. This approach seems to provide a reasonably objective methodology for the analysis of sediment-size frequency distributions. It is necessary to emphasize that although the frequency distributions may be composed of more than one component, they need not have several distinct modes. If the component means are close enough together, a distribution will be observed which has only one mode, but it may be skewed. Despite this, we will use the loose terminology of "bimodal" to convey the concept of a single frequency distribution comprised of two or more components, but not necessarily with two or more distinct modes.

Having suggested that the size distributions may be thought of as consisting of more than one component, it becomes interesting to speculate whether this seeming structure is merely a fortuitous artefact, or whether it represents a real aspect of the deposit. If two size components are present, it seems reasonable to expect that these components also may be reflected in some other characteristics of the deposit. Following Griffiths (1967), a rock specimen may be defined uniquely as

$$P = f(m, s, sh, o, p) \qquad (2)$$

where the properties (P) are a function of mineralogy (m), size (s), shape (sh), orientation (o) and packing (p). Where there is evidence that the size shows bimodal characteristics, are any of the other properties bimodal?

## EXAMPLE

To test this concept, six samples were taken from the swash-backwash zone of the beach at Dungeness, Kent. The samples were taken from the west side of this cuspate foreland, near the middle of a nearly straight section, about 6 km long. The six samples were taken at 25-m intervals, parallel to the shore line. About 40 gm of sand were collected from the top layer of the sand on the beach, over an area of about 1 sq m. The size frequency of the samples were determined by sieving at $\frac{1}{4}\phi$ intervals. The results of the sieving were modified by eliminating the contribution made by shell fragments. (These fragments made up less than 3 percent of the total distribution.)

The size-frequency distribution was analyzed by the method of nonlinear least squares (Clark and Garnett, 1974), and gave the results shown in Table 1. The analysis indicates the presence of two lognormal components in each sample, with the proportions of the components remaining reasonably constant between the samples.

Table 1. Analysis of size frequency distribution into two lognormal components.

| Site | Component 1 | | | Component 2 | | $\chi^2$ | df | $\alpha*$ |
|------|------|------|------|------|------|------|------|------|
|      | mean | s.d. | prop. | mean | s.d. |  |  |  |
| DW1 | 2.4745 | 0.3165 | 0.5713 | 2.7209 | 0.1637 | 0.32 | 1 | 0.3318 |
| DW2 | 2.4672 | 0.3406 | 0.6447 | 2.7603 | 0.1005 | 0.08 | 2 | 0.3297 |
| DW3 | 2.5380 | 0.3539 | 0.5788 | 2.7835 | 0.1232 | 0.19 | 2 | 0.3949 |
| DW4 | 2.5463 | 0.3533 | 0.6627 | 2.7737 | 0.0976 | 0.11 | 2 | 0.3798 |
| DW5 | 2.5526 | 0.3386 | 0.6555 | 2.7662 | 0.1000 | 0.65 | 2 | 0.3905 |
| DW6 | 2.6636 | 0.2846 | 0.5750 | 2.7697 | 0.0925 | 1.11 | 1 | 0.3803 |

Of the possibilities, the shape characteristics were chosen to examine more closely. Moss (1962, 1963, 1972) considered this in some detail, and was able to identify components on the basis of size and shape characteristics, but he did not relate them to the underlying lognormality of the size components.

A size grade was chosen which was well represented in the samples (2.75 to 3.00$\phi$); this was done to try to eliminate the confounding effect of size. A proportion of the grains were mounted in Canada balsam, on a glass slide. The expected proportion of the hypothetical shape components are given under the column headed $\alpha*$ in Table 1. Measurement of grain shape is a fairly routine procedure in sediment analysis. It was felt however that any shape differences which might occur were likely to be rather subtle, and that relatively simplistic measures of "a" and "b" axes, roundness and sphericity, were unlikely to yield the fine detail which might be required.

We have assumed, in common with many others, that shape information of a three-dimensional grain may be derived adequately from the two-dimensional outline of that grain. Some method is required here which will permit the grain periphery to be represented in a manner which is unique, and also tractable for some type of numerical analysis.

The analysis of closed curves like these is not restricted to sedimentary studies. Freeman (1961) suggested methods in which any arbitrary geometric shape may be encoded for further analysis, where a continuous figure can be represented in a discrete form. This is clearly of great merit in reducing the problem to manageable proportions.

## TECHNIQUES AND DATA

There seem to be at least three categories of techniques which have utility:

(1) Fourier techniques as suggested by Brill (1968), Schwarcz and Shane (1969), Ehrlich and Weinberg (1970), Granlund (1972), and Zahn and Roskies (1972). The use of Fourier models for image encoding suggests a kinship with optical methods, which in fact turns out to be a close relationship (Pincus and Dobrin, 1966; Kaye and Naylor, 1972).

(2) Slope density, introduced by Nahin (1972),(also Sklansky and Nahin, 1972; Nahin, 1974).

(3) Moments, presented by Hu (1962) and Alt (1962).

A useful review of descriptions of line and shape is given by Duda and Hart (1973). Each of these approaches has merits, but, with the exception of the "radial" Fourier method, used by Schwarcz and Shane (1969) and Ehrlich and Weinberg (1970), none of them have been used in a geological context. The method used here was the radial Fourier method, not for any known superiority, but simply because we were not aware then of the work which had been done in other fields. In fact the radial method has one major disadvantage compared with the other Fourier methods, because it cannot handle curves with substantial reentrants. However, published accounts suggest that the method preserves useful information, and has the advantage (Tilmann, 1973) that it was not critical that the maximum projective area be considered.

The mounted grains were magnified 250 times, and their outlines drawn, for about 100 to 120 grains at each site. These outlines then were digitized (Piper, 1970). The grains were digitized into between 36 and 60 points, depending on the size and complexity of the outline. The cartesian coordinates of each of these outlines were converted into polar coordinates by first determining the grain "center of gravity", and using this point as the origin for the polar coordinates. The Fourier descriptor of the outline was determined in terms of the polar coordinates, but in a manner which differed slightly to that of Ehrlich and Weinberg. It is easier to solve a Fourier series if the data points are equally spaced (in this example, at equal angular separation). Ehrlich and Weinberg use a linear interpolation scheme to provide the equal spacing, but this could introduce unwanted bias. Here equal spacing was achieved by first fitting a bicubic spline (Ahlberg, Nilson, and Walsh, 1967) to the grain periphery. A spline has the property of passing through all the data points, as a smooth curve. New data points at equal angular increments were calculated on the basis of the spline (Fig. 1). A Fourier series then was fitted to the new points.

The Fourier series may be expressed as

$$r(\theta) = a_0/2 + \sum_{i=1}^{\infty} a_i \cos(i\theta - \phi_i)$$

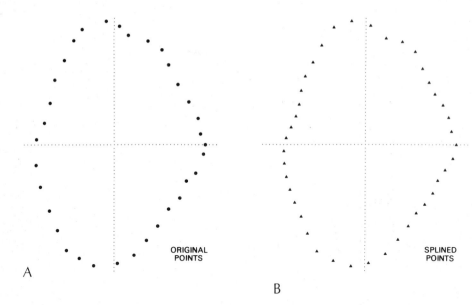

Figure 1. *A*, original data points outlining grain periphery; and *B*, equally spaced points calculated from bicubic spline fit.

where r is the radius at any given angle $\theta$,
  $a_i$ represents the contribution of the $i$th harmonic, and
  $\phi_i$ represents the phase angle (offset) of that harmonic.

This form of the expression would describe a continuous periphery. Because the periphery is not continuous in this example, but quantized, the series is truncated to n/2 terms, where n is the number of data points. In fact, in this application, the series was truncated further, to only eight terms. The Fourier equation therefore becomes

$$r(\theta) = a_0/2 + \sum_{i=1}^{8} a_i \cos(i\theta - \phi_i) \qquad (3)$$

In order to standardize the $a_i$ terms to a size-independent form, they were each divided by the average radius term ($a_0/2$). The eight terms of the truncated Fourier series retain about 85 to 90 percent of the information contained in the quantized curve (Fig. 2). A typical line spectrum is given in Figure 3. The $a_i$ terms (the harmonic amplitudes) have the convenient property of being origin independent (or rotation invariant), which allows the amplitudes from one grain to be compared with those from another. The phase angles are clearly not rotation invariant, and therefore were dropped from the subsequent analysis.

The procedure adopted is a fairly standard one in pattern recognition. Meisel (1972) outlined the methodology as one which

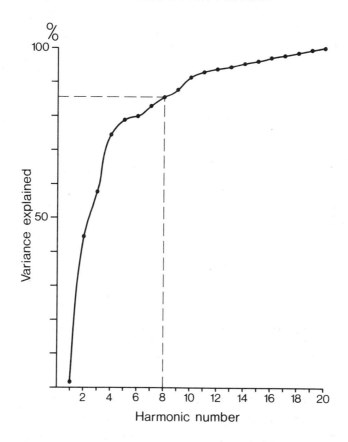

Figure 2. Typical relationship of variance explained (in percent) by Fourier series fit (harmonic number). Almost 90 percent is explained by truncated series at eight terms.

proceeds from the physical system (the sand grains), to the measurement space (the shape descriptors), into pattern space and "reduced" pattern space (the truncated Fourier series), and from there into some type of clustering procedure, from which a decision rule may be constructed in order to classify other data points (Fig. 4). The groupings themselves also may be used to summarize or exhibit the data.

Attention must be given to the clustering or grouping techniques, whereby naturally occurring homogeneous groups are determined in the data, remembering that it is anticipated that these groups may be present in the proportions as given in Table 1. Many of the classical clustering techniques suffer from the drawback that they are not able to handle large data sets. A total of 713 samples, with eight variables, is not an intolerably large data set, but simplistic number crunching perhaps is not the most subtle or rewarding technique to employ. With this in mind, each

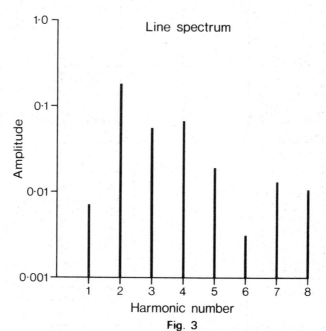

Figure 3. Typical line spectrum from Fourier fit.

of the six sites was analyzed individually. Site one (DW1) was used as a type of training set, where some conclusion was drawn about the nature of the samples. These conclusions then were tested on the other sites. This permits the consistency of the conclusions to be evaluated.

Some limitations of the classical clustering techniques are summarized by Howarth (1973). To avoid many of the usual drawbacks of clustering three techniques were employed. Nonlinear mapping (Sammon, 1969, 1970) was introduced into geology by Howarth (1973). Nonlinear mapping (NLM) is a method in which a multidimensional situation is represented in fewer dimensions (commonly, but not necessarily, two), with a minimum amount of induced distortion. The rationale of the approach is that the human eye (together with the human brain) is better able to distinguish groups than any inflexible algorithm.

The nonlinear maps, however, proved to be of limited value in this instance. The map of the first site is given in Figure 5. No groups are readily apparent. Table 2 gives the error present in the mapping, together with the probable dimensionality. The maps suggest one of two things; either there are no groups, or they are overlapping to a fairly high degree. Given the fairly high error present, it is perhaps not surprising that clusters are not observed. The mapping of all 713 individuals indicated no grouping either. This was somewhat encouraging, because it suggests that there was no "drift" or change in the shape charac-

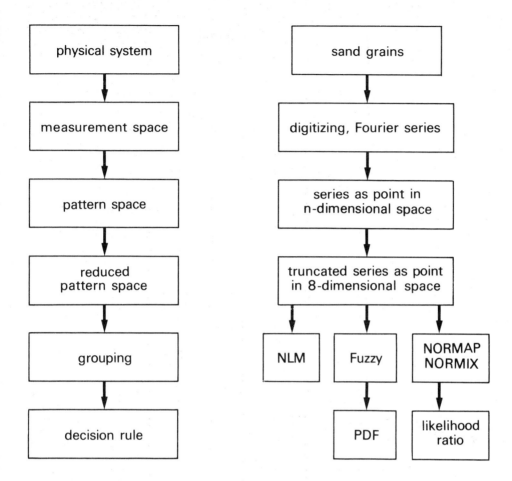

Figure 4. Standard procedure in pattern recognition (after Meisel, 1972).

teristics between the six sites (Fig. 6).

Fuzzy-set analysis also was used to seek out the groups (Zadeh, 1965). An example of a fuzzy set (Gitman and Levine, 1970) is the set defined as "all the very tall buildings", thus it is possible to see that it is a class of objects with a continuum of grades of membership. The algorithm of Gitman and Levine (1970) will detect unimodal fuzzy sets, and as such, will detect concentrations of points which may have irregular shapes (Fig. 7). A threshold parameter is used, whose value is somewhat arbitrary; different values of the threshold parameter can give different numbers of groups (and perhaps different groups). An example of how the number of groups may differ with the threshold is given in Table 3. Site one was analyzed extensively, with the object of determining those threshold values which gave two main clusters with approximately the expected proportion of members. Eight such values were determined, and are given in Table 4, together

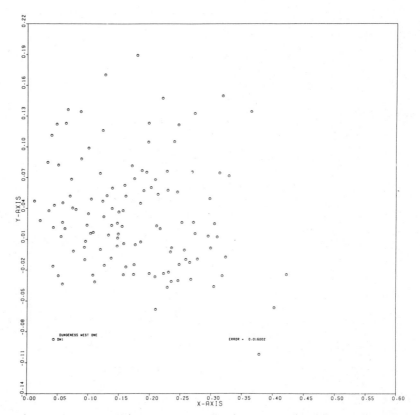

Figure 5. Nonlinear map of site one (DW1).

with the group sizes. These eight groupings were examined closely to derive consistently appearing groups. This provided three groups, one of 46 members which made up group 1, one of 55 members making up group 2, and a further 20 members which were unclassified, because they did not occur in the two core groups with regularity.

The other five sites were analyzed with the eight thresholds, and provide the results in Table 4. Although the results are not as decisive as might have been hoped, they do indicate the possible

Table 2. Error on nonlinear mapping.

| Site | mapping error (%) | probable dimensionality |
|------|-------------------|-------------------------|
| DW1  | 15.845            | 2                       |
| DW2  | 16.604            | 2                       |
| DW3  | 17.243            | 2                       |
| DW4  | 19.939            | 2                       |
| DW5  | 23.128            | 2                       |
| DW6  | 15.710            | 2                       |

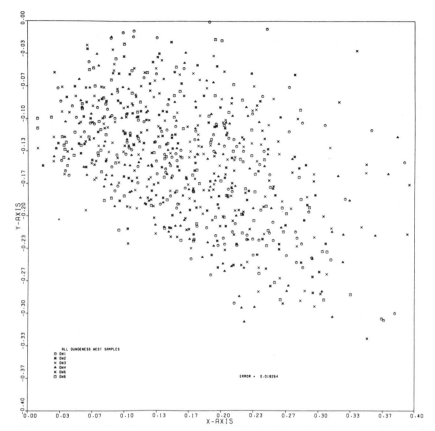

Figure 6. Nonlinear map of 713 individuals at all six sites.

presence of two major clusters at most of the sites. In interpreting the results, it is probably wise to regard groups of ten or fewer members as spurious, resulting from the fact that we are dealing with a finite (sampling) situation. Gitman and Levine (1970) note that a finite sample from a Gaussian distribution can be composed of several modes.

A decision rule also was constructed, based on the two "core" groups of site one. Because these groups are likely to be rather irregular an empirical discriminant method (Howarth, 1971; Specht, 1967a, 1967b) was employed. This has the virtue of embodying no assumptions about the nature of the underlying distributions. The classification provided by this polynomial discriminant function corresponds to a fair degree with the groupings provided by the fuzzy-set analysis. The proportions of the two groups present at the six sites is given in Table 6. The twenty unclassified individuals of site one were classified by the polynomial discriminant function and added into the cores for the table. The relative consistency of the proportions again confirms the absence of drift in the shape characteristics.

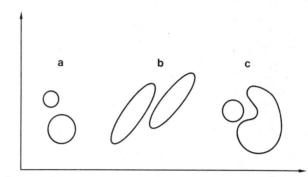

Figure 7. Shapes of point concentrations, a=circular, b=elongated, c=irregular.

The techniques used do not allow for any great amount of overlap of the components, but it seems reasonable to suggest that a high degree of overlap is present (assuming the components themselves exist). Multivariate-mixture analysis (Wolfe, 1970) permits clustering of overlapping groups. In providing this highly sophisticated analysis, the method is highly parametric. It can be seen as a multivariate extension of the methods used in analyzing the size-frequency distributions. It is assumed that the observed distribution comprises a mixture of several multivariate normal distributions. The distribution therefore is characterized by the vector of means, the covariance matrix, and the proportion, for each of the components. This requires the estimation of a large number of parameters. In an effort to reduce the computational effort, an alternative is given by Wolfe. Instead of allowing the covariance matrices to be unconstrained, the alternative requires that the covariance matrix for each of the components is equal. This reduces the number of parameters considerably. Wolfe terms the unconstrained solution NORMIX, and the

Table 3. Results with different thresholds for fuzzy-set analysis at site DW1.

| Threshold value | no. of groups | size of each group |
|---|---|---|
| 0.0069 | 4 | 8   71   32   10 |
| 0.0088 | 3 | 12   105   4 |
| 0.0107 | 2 | 109   12 |
| 0.0126 | 3 | 104   12   5 |
| 0.0145 | 3 | 102   7   12 |
| 0.0164 | 4 | 102   7   11   1 |
| 0.0183 | 3 | 107   11   3 |
| 0.0202 | 4 | 113   1   6   1 |
| 0.0221 | 9 | 1   7   85   7   6   1   10   1   3 |
| 0.0240 | 5 | 91   12   7   10   1 |
| 0.0259 | 4 | 111   6   1   3 |
| 0.0278 | 6 | 7   49   52   9   1   3 |

Table 4. Selected thresholds, with group sizes from fuzzy-set analysis, for all sites. Groups with 10 or fewer members have been omitted.

| threshold | DW1 | DW2 | DW3 | DW4 | DW5 | DW6 |
|---|---|---|---|---|---|---|
| 0.0069 | 71 32 | 106 12 | 117 | 86 35 | 73 30 | 86 16 |
| 0.0278 | 49 52 | 104 12 | 81 33 | 96 15 | 53 15 28 21 | 58 14 40 |
| 0.0354 | 72 28 | 99 12 | 117 | 56 31 | 65 49 | 107 |
| 0.0373 | 69 31 | 102 12 | 117 | 67 22 21 | 91 17 | 99 |
| 0.0411 | 40 57 | 105 12 | | 17 49 33 | 77 39 | 13 62 22 |
| 0.0525 | 54 49 | 115 | 117 | 103 | 79 38 | 104 |
| 0.0544 | 67 40 | 75 48 | 105 11 | 78 27 | 117 | 109 |
| 0.0563 | 50 66 | 65 58 | | 82 26 | | 86 11 11 |

constrained NORMAP. Both forms of the analysis were used. Clearly, there is no a priori reason to suppose that the shape characteristics should correspond to a multivariate normal distribution of the form required by the analysis; there is no reason to suggest that it should not. The marginal distributions for the amplitudes of harmonic 2 and 3 are given in Figure 8. These two variables are the two with maximum variance in every situation. There is some evidence on the basis of these marginal distributions for suspecting bimodality. A peculiarity of the variables is that they are bounded; the lowest value possible for an amplitude is zero, and the highest (by the definition used here) is unity.

It is possible to test the results of the multivariate-mixture analysis to some extent. The optimum number of components can be established by testing the hypothesis that there is one type, two types, three types, etc. Given the fairly low number of individuals present, and the large number of parameters to be estimated, it was indeed unwise to proceed to more than three types, or components. The hypothesis testing is give in Table 5. It can be seen that for the NORMAP analysis (with equal covariance matrices), the favored solution tends to be a three-component solution, whereas for the NORMIX analysis, it is a two-component solution. This may be explained with reference to Figure 9, where it is suggested

Table 5. Results from multivariate-mixture analysis, giving selected number of components.

| Site | Chosen number of components | | |
|---|---|---|---|
| | NORMAP | NORMIX | NORMAP3/NORMIX2 |
| DW1 | 3 | 2 | 2 |
| DW2 | 3 | 2 | 2 |
| DW3 | 2 | * | 3 |
| DW4 | 2 | * | * |
| DW5 | 3 | 2 | 2 |
| DW6 | 3 | 2 | 2 |

*no solution possible

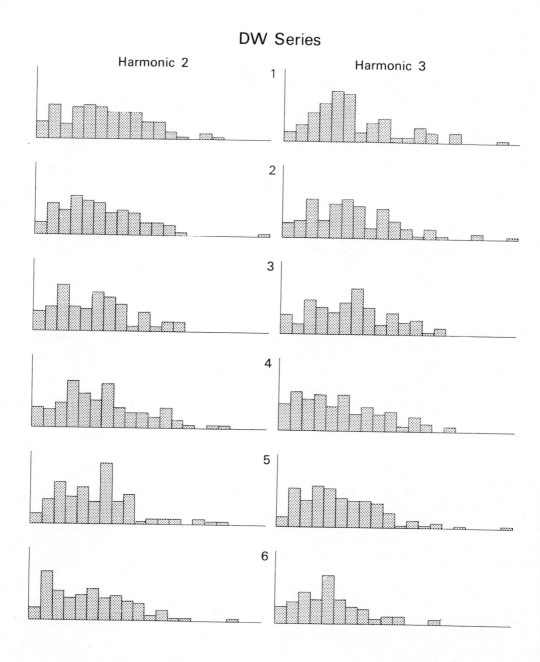

Figure 8A. Distribution of amplitudes (from 0 to 1) of harmonic 2 and harmonic 3 for all sites (DW1 through DW6). These are maximum variance directions.

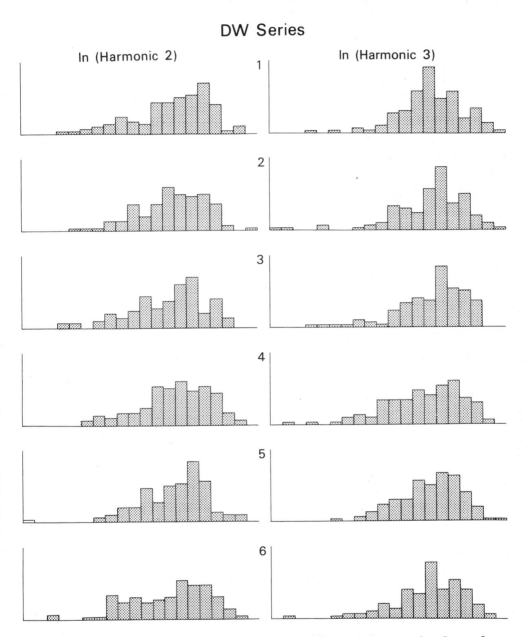

Figure 8B. Distribution of natural logarithms of amplitudes of harmonic 2 and harmonic 3 for all sites. Because this transformation does not seem to normalize distributions, it is not used in analysis.

**NORMAP**
**3 components**

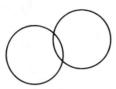

**NORMIX**
**2 components**

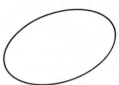

Figure 9. Graphic representation of NORMAP analysis (with equal covariance matrices) which result in three-component solution, and NORMIX analysis which tends toward two-component solution.

that two of the equal covariance matrices are attempting to approximate the single larger covariance matrix in the NORMIX analysis. Testing the NORMAP 3-type results against NORMIX 2-type results tends to confirm this view. In Table 6, where the proportions expected on the basis of the size analysis are compared, the proportions are those derived from the NORMIX analysis, except in the situation of DW3 and DW4, where the NORMAP results were used, because the tests suggest that these were the better choice of solution.

## RESULTS

The results (Table 6) are in fair agreement although derived by three different methods. The poorer performance of the multi-

Table 6. Larger proportions of shape components derived from size analysis, polynomial discriminant function based on cores from fuzzy-set analysis (PDF), and multivariate-mixture analysis (MM).

| Site | $1-\alpha^*$ | PDF | MM |
|------|--------|--------|--------|
| DW1 | 0.6682 | 0.5702 | 0.6502 |
| DW2 | 0.6703 | 0.6239 | 0.5642 |
| DW3 | 0.6051 | 0.6410 | 0.8000 |
| DW4 | 0.6202 | 0.6160 | 0.9600 |
| DW5 | 0.6095 | 0.6016 | 0.7925 |
| DW6 | 0.6197 | 0.6429 | 0.6338 |

variate-mixture analysis probably can be attributed to the problems associated in estimating the covariance matrices from a data set which was on the small side (Ball, 1965). These results would seem to confirm the view that there exist, in the sediments analyzed, two lognormal size components which are reflected in the shape characteristics of the sediment. We intend extending the analysis both to consider the other size ranges within the same deposits, and to consider other sites.

How may these shape components be explained? Two possibilities seem attractive. The two-shape components may be either the result of different transport mechanisms, or be inherited characteristics.

The results obtained by Kolmer (1973) tend to support the concept of two transport mechanisms. He suggested the presence of two saltation populations, one associated with the swash and the other with the backwash. The results in Table 1 may be interpreted in this light easily. The coarser component may be deposited in the swash. The backwash may be lower in transporting power, due to the return of some of the water as percolation. This may account for the finer component. The slightly less well-sorted nature of the coarser component may be related to the higher degree of turbulence in the swash. Waddell (1973) indicated that the sand arrived on the beach "as suspended load entrained in the uprush. The subsequent downslope movement of this material occurred as bed load in the backwash". This again may relate to the two size components. The suspended load may tend to favor the more angular grains, whereas in the backwash the more rollable grains may move more easily. Morris (1957) indicated that the roundness of grains is related in a rather complex manner to fluid velocity.

As an alternative explanation, component one may be derived from one environment (e.g. a river), whereas the other may be derived from another (e.g. offshore). This could account similarly for the two shape and size components.

The work presented here suggests that the lognormal components determined in size analysis are real features, and are reflected in the shape characteristics. The actual mechanisms giving rise to these shape and size components are not clear.

## ACKNOWLEDGMENTS

This analysis presented here represents part of the work being carried out by M.W. Clark for the degree of Ph.D. in the University of London. Financial support for the presentation of the paper was granted by the Royal Society, the Hilary Bauerman Trust, the Department of Mining and Mineral Technology (Imperial College), and the Department of Geography (London School of Economics). Thanks also must be expressed to the Computer Units of Imperial College, Kings College and the London School of Economics. The nonlinear mapping algorithm (NLM) was developed at the Rome Air Development Center, New York. The NORMIX/NORMAP and unimodal fuzzy-set analysis programs were made available through the University of London Computer Centre.

## REFERENCES

Ahlberg, J.H., Nilson, E.N., and Walsh, J.L., 1967, The theory of splines and their application: Academic Press, New York, 284 p.

Alt, F.L., 1962, Digital pattern recognition by moments: Jour. Assoc. Comput. Mach., v. 9, no. 2, p. 240-258.

Bagnold, R.A., 1941, The physics of blown sand and desert dunes: Chapman & Hall, London, 265 p.

Ball, G.H., 1965, Data analysis in the social sciences: What about the details?: AFIPS Conf. Proc. v. 27, no. 1, p. 533-559.

Brill, E.L., 1968, Character recognition via Fourier descriptors: WESCON Tech. Papers, Ses. 25 (Qualitative pattern recognition through image shaping), p. 1-10.

Clark, I., and Garnett, R., 1974, The detection of multiple mineralisation phases by statistical methods: Trans. Inst. Min. Metall., v. 83, no. 809, p. A43-A52.

Clark, M.W., 1976, A discussion of some methods for the statistical analysis of multimodal distributions, and their application to grain size data: Jour. Math. Geology, v. 8, in press.

Curray, J.R., 1960, Tracing sediment masses by grain size modes: Rept. 21st Sess. Intern. Geol. Congress (Norden), pt. 23, p. 119-130.

Doeglas, D.J., 1946, Interpretation of the results of mechanical analyses: Jour. Sed. Pet., v. 16, no. 1, p. 19-40.

Duda, R.O., and Hart, P.E., 1973, Pattern classification and scene analysis: Wiley Interscience, New York, 502 p.

Ehrlich, R., Orzeck, J.J., and Weinberg, B., 1974, Detrital quartz as a natural tracer - Fourier grain shape analysis: Jour. Sed. Pet., v. 44, no. 1, p. 145-150.

Ehrlich, R., and Weinberg, B., 1970, An exact method for characterization of grain shape: Jour. Sed. Pet., v. 40, no. 1, p. 205-212.

Folk, R.L., 1971, Longitudinal dunes of the northwestern edge of the Simpson Desert, Northwestern Territory, Australia, 1. Geomorphology and grain size relationships: Sedimentology, v. 16, no. 1, p. 5-54.

Freeman, H., 1961, On the encoding of arbitrary geometric configurations: IRE Trans., Elec. Comp., EC-10, no. 2, p. 260-268.

Gitman, I., and Levine, M.D., 1970, An algorithm for detecting unimodal fuzzy sets and its application as a clustering technique: IEEE Trans. Comp., C-19, no. 7, p. 583-593.

Granlund, G.H., 1972, Fourier preprocessing for hand print character recognition: IEEE Trans. Comp., C-21, no. 2, p. 195-201.

Griffiths, J.C., 1967, Scientific method in analysis of sediments: McGraw-Hill Book Co., New York, 508 p.

Howarth, R.J., 1971, An empirical discriminant method applied to sedimentary rock classification from major element geochemistry: Jour. Math. Geology, v. 3, no. 1, p. 51-60.

Howarth, R.J., 1973, Preliminary assessment of a nonlinear mapping algorithm in geological context: Jour. Math. Geology, v. 5, no. 1, p. 39-57.

Hu, M-K., 1962, Visual pattern recognition by moment invariants: IRE Trans., Inf. Theory, IT-8, no. 2, p. 179-187.

Jones, T.A., 1969, Determination of 'n' in weight frequency data: Jour. Sed. Pet., v. 39, no. 4, p. 1473-1476.

Kaye, B.H., and Naylor, A.G., 1972, An optical information procedure for characterizing the shape of fine particle images: Pattern Recognition, v. 4, no. 2, p. 195-199.

Kittleman, L.R., 1964, Application of Rosin's distribution in size-frequency analysis of clastic rocks: Jour. Sed. Pet., v. 34, no. 3, p. 483-502.

Kolmer, J.R., 1973, A wave tank analysis of beach foreshore grain size distribution: Jour. Sed. Pet., v. 43, no. 1, p. 200-204.

Krumbein, W.C., and Jones T.A., 1970, The influence of areal trends on correlations between sedimentary parameters: Jour. Sed. Pet., v. 40, no. 2, p. 656-685.

McKinney, T.F., and Friedman, G.M., 1970, Continental shelf sediments of Long Island, New York: Jour. Sed. Pet., v. 40, no. 1, p. 213-248.

Mahmood, K., 1973, Lognormal distribution of particulate matter: Jour. Sed. Pet., v. 43, no. 4, p. 1161-1166.

Meisel, W.S., 1972, Computer-oriented approaches to pattern recognition: Academic Press, New York, 250 p.

Middleton, G.V., 1970, Generation of the log-normal frequency distribution in sediments, *in* Topics in mathematical geology: Consultants Bur., New York, p. 34-42.

Morris, W.J., 1957, Effects of sphericity, roundness and velocity on traction transportation of sand grains: Jour. Sed. Pet., v. 27, no. 1, p. 27-31.

Moss, A.J., 1962, The physical nature of common sandy and pebbly deposits. part I: Am. Jour. Sci., v. 260, no. 5, p. 337-373.

Moss, A.J., 1963, The physical nature of common sandy and pebbly deposits. part II: Am. Jour. Sci., v. 261, no. 4, p. 297-343.

Moss, A.J., 1972, Bed-load sediments: Sedimentology, v. 18, nos. 3/4, p. 159-219.

Nahin, P.J., 1972, A parallel machine for describing and classifying silhouettes: unpubl. doctoral dissertation, Univ. California (Irvine), 184 p.

Nahin, P.J., 1974, The theory and measurement of a silhouette descriptor for image pre-processing and recognition: Pattern Recognition, v. 6, no. 2, p. 85-95.

Pincus, H.J., and Dobrin, M.B., 1966, Geological applications of optical data processing: Jour. Geophysical Res., v. 71, no. 20, p. 4861-4869.

Piper, D.J.W., 1970, The use of the D-Mac pencil follower in routine determinations of sedimentary parameters, *in* Data processing in biology and geology: Academic Press, New York, p. 97-103.

Sammon, J.W., 1969, A non-linear mapping for data structure analysis: IEEE Trans. Comp., C-18, no. 5, p. 401-409.

Sammon, J.W., 1970, Interactive pattern analysis and classification: IEEE Trans. Comp., C-19, no. 7, p. 594-616.

Schwarcz, H.P., and Shane, K.C., 1969, Measurement of particle shape by Fourier analysis: Sedimentology, v. 13, nos. 3/4, p. 213-231.

Sklansky, J., and Nahin, P.J., 1972, A parallel mechanism for describing silhouettes: IEEE Trans. Comp., C-21, no. 11, p. 1233-1239.

Specht, D.F., 1967a, Generation of polynomial discriminant functions for pattern recognition: IEEE Trans. Elec. Comp., EC-16, no. 3, p. 308-319.

Specht, D.F., 1967b, Vectorcardiographic diagnosis using the polynomial discriminant method of pattern recognition: IEEE Trans. Bio-med. Eng., BME-14, no. 2, p. 90-95.

Spencer, D.W., 1963, The interpretation of grain size distribution curves of clastic sediments: Jour. Sed. Pet., v. 33, no. 1, p. 180-190.

Tanner, W.F., 1958, The zig-zag nature of type I and type IV curves: Jour. Sed. Pet., v. 28, no. 3, p. 372-375.

Tanner, W.F., 1964, Modification of sediment size distribution: Jour. Sed. Pet., v. 34, no. 1, p. 156-164.

Tilmann, S.E., 1973, The effect of grain orientation on Fourier shape analysis: Jour. Sed. Pet., v. 43, no. 3, p. 867-869.

van Andel, T.H., 1973, Texture and dispersal of sediments in the Panama Basin: Geol. Soc. America Bull., v. 81, no. 4, p. 434-457.

Visher, G.S., 1969, Grain size distributions and depositional processes: Jour. Sed. Pet., v. 39, no. 3, p. 1074-1106.

Waddell, E., 1973, Dynamics of swash and implication to beach response: Coastal Studies Inst., Tech. Rept. 139, Louisiana State Univ., Baton Rouge, 49 p.

Walger, E., 1961, Grain size distribution within single arenaceous beds and their genetic meaning (in German): Geologische Randschau, v. 51, no. 2, p. 494-507.

Wolfe, J.H., 1970, Pattern clustering by multivariate mixture analysis: Multivariate Behav. Res., v. 5, p. 329-350.

Zadeh, C.T., 1965, Fuzzy sets: Information and Control, v. 8, no. 3, p. 338-353.

Zahn, C.T., and Roskies, R.Z., 1972, Fourier descriptors for plane closed curves: IEEE Trans. Comp., C-21, no. 3, p. 269-281.

# MULTIDIMENSIONAL SCALING OF SEDIMENTARY ROCK DESCRIPTORS

J.H. Doveton

*Kansas Geological Survey*

## ABSTRACT

Most attributes observed from sandstones and sand-depositional environments are measured on nominal and ordinal scales, which may invalidate the strict application of conventional metric analytical methods. Nonmetric multidimensional scaling is an appropriate technique for low-dimensional ordination of nonmetric data because the method is constrained by the ordinal structure of interpoint distances rather than their metric values. The application of nonmetric MDS to two diverse examples is discussed: the scaling of the petrography of a suite of North American sandstones illustrates relationships between textural and mineralogical maturity; a two-dimensional ordination of the facies characteristics of sand sedimentation environments strongly reflects geographic disposition and hydrodynamic controls. KEY WORDS: *classification, multidimensional scaling, numerical analysis, sandstones, sedimentology.*

## INTRODUCTION

The condensation of multivariate observational data to a smaller number of composite variables is a prime objective in mathematical geology and follows the basic scientific philosophy of parsimony. These variables may be identified with fundamental causal factors either by empirical interpretation or, where warranted, by statistical hypothesis tests. Whereas, there is almost no limit to the mathematical transformation of observational data, the constraints on analysis are dictated by the distribution types, and measurement scales of the raw data.

Interval- and ratio-scale measurements are made in continuous metric space and methods for their analysis are well grounded in the main body of statistical theory. However, nominal- and ordinal-scale measurements have a more restricted information content and appropriate multivariate analysis procedures are equivocal and subject to debate. This point is particularly well illustrated in many paleontologic and paleoecologic studies where basic data commonly are in binary (presence/absence) form. Because the objective is commonly the classification of objects as distinctive entities rather than as anonymous members of large populations, a Q-mode analysis may be made, based on computations of similarity measures between objects. The object-pair similarities are used as ersatz metric data in factor analysis, cluster analysis, and related display techniques.

The Pearsonian correlation coefficient is well established on theoretical grounds as an optimal measure of similarity between variables for normally distributed continuous data, but no such external criteria apply to measures of similarity between objects. Cheetham and Hazel (1969) list 22 similarity measures used in the literature, including those perennial favorites, the Jaccard and Dice coefficients. Selection of an appropriate coefficient is based generally on the researcher's assessment of the material at hand, which may result in the infusion of genetic preconceptions prior to analysis. The range of possible measures is a reflection of the basic data itself. Because the observations are discrete, objects can at best be ranked on any variable, so that continuous measures of distances between objects can only be informal estimates rather than strictly metric quantities. Analysis of the raw data in R-mode space provides no escape from this constraint because the discrete character of the variables precludes formal parametric statistical analysis.

The most appropriate methods for multivariate analysis and display of discretely characterized data must be directed towards the *ranking* of objects in multidimensional space rather than metric transformations of their continuous distances. Although solutions to some of the metric clustering algorithms are invariant under monotonic transformations of the input similarity coefficients (Sokal, 1974), ordination by nonmetric multidimensional scaling is both true to the spirit of the original scale of measurement and generally provides a more faithful representation of the original similarity matrix.

Analogous problems arise with material derived from the study of sedimentary rocks, including sandstones. A comprehensive suite of observations made on a typical sample of sandstone will include data measured on nominal, ordinal, interval and ratio scales, as illustrated in Table 1. The investigator is faced with the choice of either conducting a piecemeal statistical analysis of separate data items, segregated according to measurement scale, or of using a comprehensive procedure geared to the most degraded measurement scale. Sandstone properties are the product of a complex of genetic processes which are continuous, rather than discrete, in nature. If sandstone samples could be represented in a continuous space based on diagnostic observable properties, the spatial configuration would be a latent reflection of the causal factors of variation. Multidimensional scaling (MDS) provides the most viable method of computing a multidimensional ordination of sandstone samples which simultaneously allows utilization of the greatest range of observational data without violating constraints imposed by their measurement scales.

Table 1. Measurement scales of sedimentary rock properties

| Measurement Scale | Observational properties |
|---|---|
| Nominal | Presence or absence of mineral species, fossils, hydraulic and biogenic structures, etc. |
| Ordinal | Ordered qualitative categories such as sorting and roundness. |
| Interval | Metric properties with no absolute zero, such as phi-scale grain-size measurements and radiometric ages. |
| Ratio | Metric properties with an absolute zero, including arithmetic-scale grain size, porosity, permeability, dimensional measures of sedimentary body geometry, and quantitative estimates of content of minerals, fossils, etc. |

## NONMETRIC MULTIDIMENSIONAL SCALING

Nonmetric MDS methods have been developed by many authors, including Torgerson (1952), Shepard (1962), Kruskal (1964), Guttman (1968) and Young (1968), utilizing a variety of approaches and viewpoints. The central philosophy of all their work is that any set of proximity measures between objects implies a spatial structure within which the objects are embedded. If the observed distances are real (continuous metric measurements), then a spatial representation must be as true as possible to the interpoint distances and the spatial recovery algorithm is designed as a nonlinear metric procedure. This type of analysis is known as "nonlinear mapping" (NLM) and is an alternative to principal component analysis for condensing the dimensionality of data (Howarth, 1973). This is especially true where a linear model is only weakly appropriate.

Where the proximity measures are implied distances, and therefore nonmetric, the spatial recovery algorithm is geared to a monotonic transformation, with a translation from implicit distances to explicit distances between study objects. The recovered spatial coordinates are intrinsically arbitrary and may be reflected or rotated because the crux of the representation is contained in the interpoint distances and the consequent multidimensional rankings of the objects. However, if the observations are controlled by latent causes which can be represented in a spatial framework, axes may be recognized in the MDS ordination that can be identified with genetic factors.

MDS algorithms are iterative procedures in which the spatial coordinates of object points are successively modified in multidimensional space so that the Euclidean distances between them conform more closely to their measured similarities. At each iteration, a monotonic regression is made between spatial explicit distances and their corresponding similarity measures. The resulting disparities ($\hat{d}_{ij}$) are compared with the explicit distances ($d_{ij}$) by Kruskal's (1964) "stress" measure where

$$\text{stress} = \sqrt{\frac{\Sigma(d_{ij}-\hat{d}_{ij})^2}{\Sigma(d_{ij}-\bar{d})^2}}$$

If the explicit distances are a monotonic transformation of the proximity measures, there is a perfect spatial representation of the objects in terms of rank order, and the stress value is zero. (Stress is conceptually similar to variance about a regression line). At an intermediate iteration, the stress values are larger than zero, and the disparities are used to compute corrective movements of points to the configureation for the next iteration, when the stress value will be less. The corrections (designated "c" by Young) are given by the formula (Petersen and Jensen-Butler 1973):

$$c_{i\ell} = \frac{1}{n} \sum_{j \neq i}^{n} \frac{(d_{ij}-\hat{d}_{ij})(\chi_{j\ell}-\chi_{i\ell})}{d_{ij}}$$

where  $c_{i\ell}$ = correction for point i on axis $\ell$;

$\chi_{j\ell}$ = coordinate of $\chi_j$ on axis $\ell$;

n = number of object points.

In practice, the algorithm is applied to a dimensional framework whose order is set by the investigator. Initial coordinates of objects are either set arbitrarily or by the semimetric method of Young and Torgerson (1967), which defines an initial configuration from a scalar product transformation of the original proximity data. Monotonic regression then is performed, stress computed, corrections calculated and applied to the object coordinates, and the procedure is repeated iteratively. Unless the proximity measures can be monotonically transformed to explicit distances in the chosen dimensional space, the algorithm ideally converges to a stable configuration in which stress is minimized. A variety of alternate runs may be required to establish whether this minimum stress configuration represents a global or a local minimum.

The final stress figure gives an empirical idea of how well the objects can be represented in the number of dimensions selected. A table of stress-value characterizations is given by Kruskal (1964). The upper limit of the number of dimensions is set by the number of objects and their interrelationships. In order to recover the intrinsic dimensionality of the objects' spatial representation, a series of MDS runs must be made specifying different dimensions. A plot of minimum stress versus number of dimensions may be interpreted in an analogous method to principal component eigenvalues, with the exclusion of ancillary higher dimensions that do not contribute significantly to representation or the location of an "elbow" in the stress/dimension curve.

Nonmetric MDS is applied widely in the fields of psychology, geography, archaeology and biology with satisfactory results. Published studies in geology are restricted mainly to paleontological work (Rowell, McBride, and Palmer, 1973; Whittington and Hughes, 1972).

## MULTIDIMENSIONAL SCALING OF SANDSTONE PETROGRAPHIC PROPERTIES

A suite of North American reservoir sandstones collected for study under API Project 131 were sampled for a representative subset of 23 core specimens for use in this study. As shown in Table 2, the subset is drawn from a wide stratigraphic and geographic

Table 2. Sample suite of reservoir sandstones.

| Sample number | Stratigraphic Unit | Age | Classification |
|---|---|---|---|
| 1 | Basal Eau Claire | Cambrian | Orthoquartzite |
| 2 | Mount Simon Sandstone | Cambrian | Orthoquartzite |
| 3 | Green River Formation | Eocene | Arkose |
| 4 | Muddy Sand | Cretaceous | Orthoquartzite |
| 5 | Upper Morrow Formation | Pennsylvanian | Orthoquartzite |
| 6 | Lower Morrow Formation | Pennsylvanian | Orthoquartzite |
| 7 | Bucella Zone | Miocene | Arkose |
| 8 | Reagan Sandstone | Cambrian | Orthoquartzite |
| 9 | Bartlesville Sandstone | Pennsylvanian | Orthoquartzite |
| 10 | Peru Sandstone | Devonian | Orthoquartzite |
| 11 | Torpedo Sandstone | Pennsylvanian | Orthoquartzite |
| 12 | Granite Wash Sandstone | Devonian | Arkose |
| 13 | Wall Creek Sandstone | Cretaceous | Orthoquartzite |
| 14 | Lakota Sandstone | Cretaceous | Orthoquartzite |
| 15 | Bandera Sandstone | Pennsylvanian | Orthoquartzite |
| 16 | Berea Sandstone | Mississippian | Arkose |
| 17 | Boise Sandstone | Pliocene | Arkose |
| 18 | Garden Gulch Sandstone | Eocene | Orthoquartzite |
| 19 | Big Clifty Sandstone | Mississippian | Orthoquartzite |
| 20 | Clear Creek Sandstone | Pennsylvanian | Orthoquartzite |
| 21 | Bromide Sandstone | Ordovician | Orthoquartzite |
| 22 | Noxie Sandstone | Pennsylvanian | Arkose |
| 23 | Mansfield Sandstone | Pennsylvanian | Orthoquartzite |

distribution, and the samples exhibit a broad variation in textural properties and mineralogical composition. Petrographic descriptions were made from thin sections of each core sample by Hoffman (1973), and the attributes listed in Table 3 are the basis for proximity measures used as input for multidimensional scaling.

Texts on sandstone petrography such as Pettijohn (1957), Folk (1968), and Pettijohn, Potter, and Siever (1972), contain general statements on the relationships among textural and mineralogical properties of sandstones. These are based on both empirical observations and deductions drawn from current knowledge of the physical processes of weathering, transport and diagenesis of clastic sediments. As indicated earlier, numerical analysis of these same properties is hampered by the variety of measurement scales involved. However, a low-dimensional ordination of a composite suite of attributes can be made by multidimensional scaling. A representation having low stress might provide insights into the relationships among sandstone properties and, by implication, into genetic processes.

Almost all the petrographic attributes measured for the study can be considered to be effectively ordinal. This is strictly true for properties such as roundness and sorting. For accessory mineralogical composition, each constituent is absent in the majority of samples and, if present, the quantities are estimates subject to both operator error and the degree to which the thin section is representative of the core sample. As such the mineralogical figures are weakly ratio scaled and can be referenced more honestly to an ordinal scale. Each attribute was scaled in the ranked categories shown in Table 3. A raw similarity matrix was summed from the data in which entries corresponded to the number of times each sample pair was grouped within the same category for the 12 separate attributes. A standardized distance matrix was computed by calculating the square root of the sum of the square differences between each

Table 3. Ordinal attributes of reservoir sandstones

| Attribute | Number of categories |
|---|---|
| Grain size | 6 |
| sorting | 3 |
| Roundness | 5 |
| Feldspar | 3 |
| Igneous rock fragments | 3 |
| Muscovite | 3 |
| Biotite | 3 |
| Heavy minerals | 3 |
| Chert | 3 |
| Clay | 3 |
| Cement | 3 |
| Porosity | 5 |

sample vector, or

$$\delta_{ij} = \sqrt{\sum_{\substack{\ell \neq i \\ \ell \neq j}} (s_{i\ell} - s_{j\ell})^2}$$

where $\delta_{ij}$ = dissimilarity distance between sandstones i and j;

$s_{i\ell}$ = frequency of common attribute grouping of sandstones i and $\ell$ (raw similarity).

(Alternative proximity measures could be used as expressions of the distances between sandstone samples. However, as the MDS algorithm operates on the monotonicity of the distance measures, broadly equivalent measures should yield substantially similar spatial representations.)

A series of multidimensional scalings of the sandstone samples yielded spatial configurations whose minimal stress values are shown plotted against order of dimension in Figure 1. The simple decline in stress with higher dimension implies that the factors underlying petrographic variation between samples are a blended interplay of individual processes rather than representing a suite of distinctly separable controls. The relatively low stress of the two-dimensional ordination (typified by Kruskal, 1964, as a fair goodness-of-fit for the configuration) suggests that these factors can be expressed adequately in a low dimension. The broad scatter of sample points across the ordination (Fig. 2A, 2B) shows no distinctive clustering or systematic trends. Because the sandstones are diverse both stratigraphically and geographically, this is not surprising, and indicates that the gross evolution of sand sediments is registered in a blurred composite petrographic characterization rather than a segregation of samples into constricted fields.

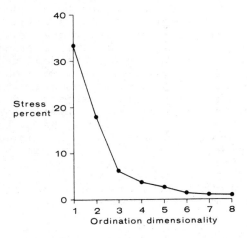

Figure 1. Stress values versus ordination dimension order for MDS of reservoir sandstones.

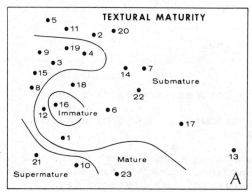 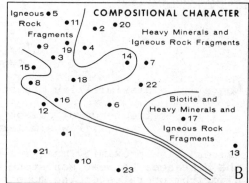

Figure 2. Two-dimensional ordination of reservoir sandstones referenced with A textural, and B compositional characters.

Each sandstone was indexed with its raw petrographic descriptors to identify ordination fields of common character and to perceive trends of possible genetic significance. A textural maturity classification based on clay content, sorting and roundness of quartz grains (after Folk, 1951) shows a distinctive arrangement of textural classes (Fig. 2A). A major trend from submature through mature to supermature across the plane is marred only by a single, immature sample (Berea Sandstone). The disruption of the pattern may be caused partly by the use of clay content as the definitive delimiter of immaturity in contrast with the grain-textural criteria used for the other classes. Moreover, the sample shows textural inversion, being mineralogically immature (10% clay, 12% feldspar) and grain texturally mature (well sorted, subangular). If reclassified as mature, the sample shows good concordance with the main trend of maturation.

Mapping of the presence, absence or trace quantity occurrence of muscovite, chert, feldspar and clay shows no clearcut arrangements capable of consistent interpretation. The wide occurrence of muscovite in a variety of depositional environments, coupled with the possibility of authigenic formation, may account for the diffuse pattern of the muscovite field. More puzzling is the failure of feldspathic sandstones to concord with textural variation. Ideally, an inverse relationship would be anticipated between the weathering and removal of primary feldspar and sand sediment maturation processes. Because the multidimensional scaling of points is merely an arrangement of the samples, with no distortion in sample properties, this problem is inherent in the raw petrographic data.

Occurrence of biotite, heavy minerals and igneous rock fragments can be combined in a composite pattern (Fig. 2B). The mineralogical compositions are matched closely by textural maturity changes and can be viewed as an implicit representation of the mainline evolution of sand fabrics from submature litharenites to supermature quartzarenites.

## MULTIDIMENSIONAL SCALING OF SAND SEDIMENTATION ENVIRONMENTS

Most current texts on sandstone morphology and petrography include an extensive treatment of modern environments of sand deposition. A major aim of these studies is the isolation of diagnostic features which may be applied to the interpretation of ancient sand bodies as a consequence of uniformitarianism. Because only rarely is a single property uniquely specific to any particular environment, paleoenvironments are recognized through consideration of a sum of characteristics or facies. These basic descriptors are referenced to a host of criteria (hydraulic, geometric, biological, compositional, etc.) and are drawn from all measurement scales, although most are nominal or ordinal in nature. Most sedimentologists integrate their observations of environmental facies patterns in a conceptual model that seems to best explain major sources of variation. These theoretical models may be given spatial descriptions in the form of charts and diagrams where sedimentary facies are related with primary axes corresponding to important environmental controls. Multidimensional scaling provides a formal numerical means of representing patterns of environmental similarities in spatial coordinates by monotonic analysis of their facies attributes.

A table from Shelton (1973) listing criteria for environmental interpretation of sand deposits was used as the basic data for computing dissimilarity distances between 20 environments (Table 4). The distances were computed as functions of the presence or absence of planar and massive bedding; small-, medium-, and large-scale cross bedding; parallel and ripple lamination; sole markings; deformational features; burrows; mudcracks; gravel; silt; sharp lower and lateral unit contacts; upward fining and upward coarsening units; clayey sideritic concretions; glauconite; allochthonous granular mineral suites; rootlet horizons; wood fragments; finely divided plant material; freshwater and brackish fauna; shallow- and deep-marine fauna; allochthonous fauna.

Multidimensional scaling resulted in a planar ordination (Fig. 3) whose low-stress value (12.8%) indicates that the environmental facies are embedded in essentially two-dimensional space. The primary axes of this space are tentatively identified with biogeographic factors and variation in hydraulic regimes. The overall pattern of the ordination shows a striking mimicry of the geographic disposition of the classified environments. The only notable exception is the placement of the deeper marine facies in a position intermediate between shallow marine and alluvial environments. Because both alluvial and turbidite sands are characterized by repetitive fining upward units with sharp lower contacts, common basal lags and allochthonous fauna, their spatial proximity is to be expected. This point may be emphasized further by a comparison of bedding sequences in Allen's (1970) ideal alluvial cycle with Bouma's (1962) ideal turbidite sequence (Fig. 4).

Table 4. Sand depositional environments (after Shelton, 1973)

| Code | Environment | Context |
|---|---|---|
| A | Aeolian dune | Continental |
| B | Piedmont | |
| C | Braided stream valley | Valley ⎫ |
| D | Meander belt valley | ⎭ Alluvial |
| E | Braided plain | |
| F | Meander belt plain | Plain |
| G | Estuarine | |
| H | Delta fringe | Deltaic |
| I | Distributary | |
| J | Regressive barrier beach bar | Coastal |
| K | Tidal pass | |
| L | Tidal-flat channel | Interdeltaic |
| M | Transgr.-stat. beach barrier bar | |
| N | Regressive nearshore | |
| O | Nearshore tidal flat | |
| P | Offshore | Shallow |
| Q | Transgressive nearshore | Marine |
| R | Channel (proximal) | |
| S | Turbidity current flow (proximal) | Deep |
| T | Basin floor (distal) | |

## CONCLUSIONS

The techniques of multidimensional scaling provide a means for low-dimensional representation of multivariate data. By virtue of the technique's analysis of monotonicity, it is one of the few methods appropriate for measurements made on nominal and ordinal scales. Because most sedimentological observations are measured on low-grade information scales, MDS offers a method of treating composite data sets and can be viewed as a purely graphic aid or as a semimetric base for further numerical analysis.

The two examples discussed are highly generalized; their principal drawback is that they are drawn from samples that come close to embracing the universal set of sandstones. More specific studies directed to the analysis of restricted suites of sandstones should yield interesting results, particularly because

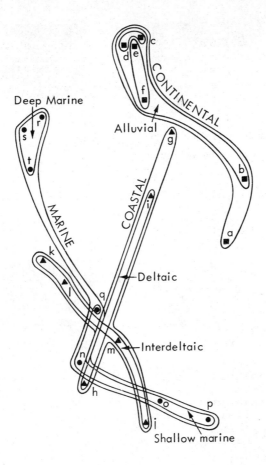

Figure 3. Two-dimensional MDS of depositional sand environments based on facies criteria. (Environments lettered according to key of Table 4.)

the nature of much of the observational data precludes the use of better known statistical techniques.

## ACKNOWLEDGMENTS

The writer is indebted to Drs. John C. Davis and Curtis D. Conley for their interest and critical reading of the manuscript. The multidimensional scaling runs were made on the University of Kansas Honeywell 635 computer using the program AARSCAL 1 written by Jensen-Butler and Petersen (1973).

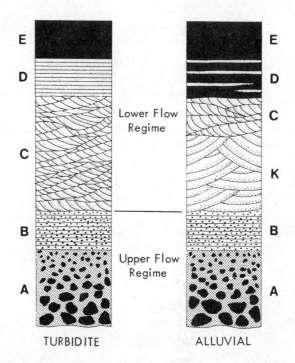

Figure 4. Idealized turbidite and alluvial bedding sequences (after Bouma and Allen). A-graded; B-laminated; C-ripple laminated; D-parallel laminated; E-pelitic interval; K-cross bedded.

## REFERENCES

Allen, J.R.L., 1970, A quantitative model of grain size and sedimentary structures in lateral deposits: Geol. Jour., v. 7, pt. 1, p. 129-146.

Bouma, A.H., 1962, Sedimentology of some flysch deposits: Elsevier Publ. Co., Amsterdam, 168 p.

Cheetham, A.H., and Hazel, J.E., 1969, Binary (presence-absence) similarity coefficients: Jour. Paleontology, v. 43, no. 5, p. 1130-1136.

Folk, R.L., 1951, Stages of textural maturity in sedimentary rocks: Jour. Sed. Pet., v. 21, no. 3, p. 127-130.

Folk, R.L., 1968, Petrology of sedimentary rocks: Texas Hemphill's Book Store, Austin, 170 p.

Guttman, L., 1968, A general non-metric technique for finding the smallest coordinate space for a configuration of points: Psychometrika, v. 33, p. 469-506.

Hoffman, D.S., 1973, A petrographic study of selected sandstones: API Project 131, Kansas Geol. Survey, unpubl. intern. rept., 37 p.

Howarth, R.J., 1973, Preliminary assessment of a nonlinear mapping algorithm in a geological context: Jour. Math. Geology, v. 5, no. 1, p. 39-57.

Jensen-Butler, C., and Petersen, V.C., 1973, AARSCAL 1, AARSCAL 2--Two non-metric multidimensional scaling programs: Computer Appl., v. 1, no. 1, p. 1-50.

Kruskal, J.B., 1964, Nonmetric multidimensional scaling: a numerical method: Psychometrika, v. 29, p. 115-129.

Petersen, V.C., and Jensen-Butler, C., 1973, An introduction to the problem of scaling and to non-metric multidimensional scaling: Arbejdsrapport nr. 1, Geografisk Inst., Aarhus Univ., Denmark, 37 p.

Pettijohn, F.J., 1957, Sedimentary rocks (2nd ed.): Harper & Bros., New York, 718 p.

Pettijohn, F.J., Potter, P.E., and Siever, R., 1972, Sand and sandstone: Springer-Verlag, New York, 618 p.

Rowell, A.J., McBride, D.J., and Palmer, A.R., 1973, Quantitative study of Trempealeavian (latest Cambrian) trilobite distribution in North America: Geol. Soc. America Bull., v. 84, no. 10, p. 3429-3442.

Shelton, J.W., 1973, Models of sand and sandstone deposits: A methodology for determining sand genesis and trend: Oklahoma Geol. Survey Bull. 118, 122 p.

Shepard, R.N., 1962, The analysis of proximities: Multidimensional scaling with an unknown distance function, I: Psychometrika, v. 27, p. 125-139.

Sokal, R.R., 1974, Classification: Purposes, principles, progress, prospects: Science, v. 185, no. 4157, p. 1115-1123.

Torgerson, W.S., 1952, Multidimensional scaling: I Theory and method: Psychometrika, v. 17, p. 401-419.

Whittington, H.B., and Hughes, C.P., 1972, Ordovician geography and faunal provinces deduced from trilobite distribution: Royal Soc. London Philos. Trans., Ser. B, v. 263, p. 235-278.

Young, F.W., 1968, TORSCA-9: A FORTRAN IV program for non-metric multidimensional scaling: Behavioral Science, v. 13, p. 343-344.

Young, F.W., and Torgerson, W.S., 1967, TORSCA, a FORTRAN IV program for Shepard-Kruskal multidimensional scaling analysis: Behavioral Science, v. 12, p. 498.

# THE IDENTIFICATION OF DISCONTINUITIES FROM AREALLY DISTRIBUTED DATA

Stephen Henley

*Institute of Geological Sciences*

### ABSTRACT

A number of approaches are used to identify discontinuities (e.g. faults) from areally distributed data. It is possible to extend into two dimensions some of the solutions for one-dimensional data, but problems arise from different topological properties of surface data; a discontinuity may be of limited extent. This property restricts the application of any method which adopts a classificatory approach.

Some simple numerical techniques are proposed in the two-dimensional situation which merely identify discontinuities with no attempt at classification of the data. A triangular or rectangular grid may be superposed over the area of interest, using data points or interpolating to regular grid points, and then compute values for gradient or rate of change of gradient between adjacent points, using a linear or nonlinear (e.g. spline) algorithm. Discontinuities are identified by masking values which do not exceed a given threshold. An alternative approach would be to concentrate on the grid triangles or rectangles and compute the angles (in three dimensions) between the surfaces defined by local trend analysis of adjacent cells. Yet another approach would involve computation of local values of variance, range, or some other measure of variability for each grid cell, and thus obtain a "local variability index" map - the variability obviously would be highest in grid cells straddling major discontinuities.

All these techniques, proposed specifically for analysis of areally distributed data containing discontinuities, are computationally simple and intuitively reasonable. KEY WORDS: *data mapping, autocorrelation, classification, cluster analysis, discontinuity analysis, structure.*

## INTRODUCTION

Little work seems to have been published on the problem of identification by computer of discontinuities in spatially distributed data. A number of solutions have been determined for the one-dimensional situation (Webster, 1973; Hawkins and Merriam, 1973, 1974, 1975; Kulinkovich, Sokhranov, and Churinova, 1966). The two-dimensional situation however has received little attention possibly because direct extension of the one-dimensional solutions to two-dimensional data does not necessarily yield meaningful solutions as has been pointed out by Webster (1973).

There are two different approaches to the identification of discontinuities in two-dimensional data; (1) data may be segmented into a set of zones which are separated by discontinuities, or (2) some technique may emphasize any discontinuities that are present with no attempt at zonation of the data.

The first approach includes such techniques as cluster analysis, and the minimal-spanning tree method (Zahn, 1971), as well as the more direct extension of the one-dimensional techniques by segmenting a series of sections across the area of interest. The second approach includes a broader range of methods drawn from the fields of multivariate statistics - multidimensional scaling (Kruskal, 1964), nonlinear mapping (Sammon, 1969), from pattern recognition, texture analysis (McCormick and Jayaramurthy, 1975), and methods which are applied now to problems of spatially distributed data such as analytical hill-shading (Yoeli, 1965).

## SEGMENTATION OF ONE-DIMENSIONAL SEQUENCE DATA

An excellent summary of the techniques which may be used to segment one-dimensional data has been presented by Hawkins and Merriam (1975). Four techniques are described:

(i) the numeric differentiation method of Kulinkovich, Sokhranov, and Churinova (1966) infers a discontinuity at any point where the gradient (measured by the successive difference $z_{i+1} - z_i$ if the data points are equally spaced) numerically exceeds a preset threshold value;

(ii) the split moving-window method of Webster (1973) uses a window which is passed across the data. This window is split at its center and the mean Z value for each half is computed. If the difference between these two means reaches a local maximum exceeding a present threshold value, a discontinuity is inferred at the position of maximum difference;

(iii) the maximum level variance method of Testerman (1962) and Hawkins and Merriam (1973) is a divisive clustering technique with a contiguity constraint, in which the entire sequence is split into the desired number k segments by choosing k - 1 boundaries

(discontinuities) to maximize the between-segment sum
of squared deviations. Because the entire sequence,
and thus the whole of each segment, is examined before
allocation of boundaries, it is possible for this technique to ensure homogeneity of each segment; and

(iv) the piecewise-regression method (Hawkins, 1972)
is similar to the maximum level-variance method except
that each segment is fitted by a low-order polynomial
or other function and the segment boundaries chosen to
minimize the total residual variance. Pavlidis (1972)
has shown that it is possible to extend this method
to two-dimensional data, but also has pointed out that
the approach is not suitable for all types of data.

## PROPERTIES OF SURFACES

The extension of discontinuity identification methods to two
dimensions is not as simple as it might seem. In a one-dimensional
sequence of z values, any discontinuity automatically partitions
the data into two sets, or segments (Fig. 1). The two-dimensional
equivalent is the partitioning of an area into a number of poly-
gonal zones (Fig. 2). However, it is easy to construct more complex patterns, in which the zones are not simple polygons (Fig. 3)
or in which a discontinuity does not partition the data (Fig. 4).
The topology of two-dimensional space is more complex than that
of one-dimensional space, and it is obvious that a segmentation
approach is not always appropriate.

## WHAT IS A DISCONTINUITY ?

Before deciding on the appropriate method, the type of discontinuity that is to be identified must be defined. If a fault is
considered as a single discontinuity, an abrupt change in the z
value (Fig. 5A), it may be possible to use some clustering technique to locate discontinuities in the data. If, however, a
fault is considered as two discontinuities, that is two breaks in
slope (Fig. 5B), such methods are less appropriate. Similarly, if
searching for ridges and valleys (Fig. 5C) or for a single,

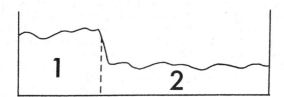

Figure 1. In one-dimensional problem,
discontinuity always partitions data into two disjoint
sets.

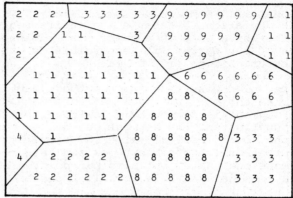

Figure 2. Direct extension of one-dimensional situation to two dimensions is possible if it is specified that discontinuities must partition data into disjoint convex sets.

isolated change of slope (Fig. 5D), analytical hill-shading and its derivatives, and pattern-recognition techniques, are more likely to be useful. These points are summarized in Table 1. It is seen that zonation (classification) techniques are of limited usefulness compared with the pattern-recognition and spatial-analysis techniques which were designed for just this type of problem.

SEGMENTATION OF TWO-DIMENSIONAL DATA

If a discontinuity is defined as a *boundary* between two zones of contrasting characteristics, so that no zone contains internal

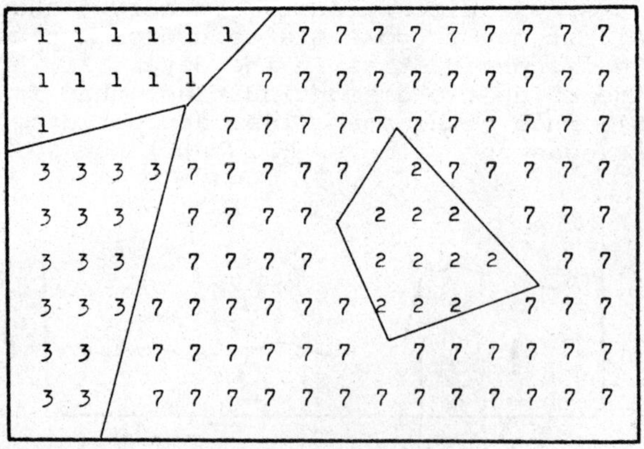

Figure 3. In two dimensions it is possible for discontinuities to define nonconvex sets.

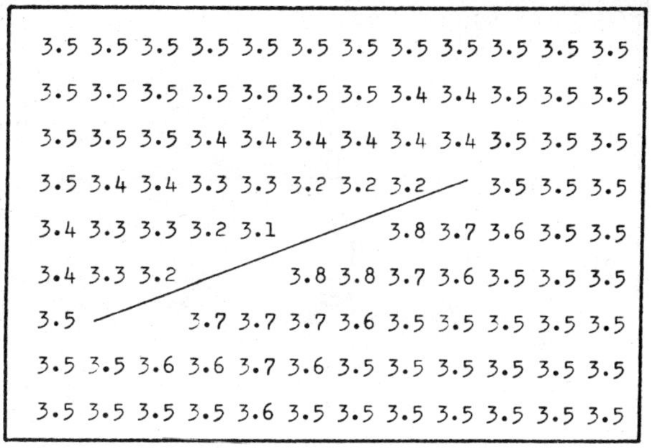

DISCONTINUITIES FROM AREALLY DISTRIBUTED DATA         161

```
3.5  3.5  3.5  3.5  3.5  3.5  3.5  3.5  3.5  3.5  3.5  3.5
3.5  3.5  3.5  3.5  3.5  3.5  3.5  3.4  3.4  3.5  3.5  3.5
3.5  3.5  3.5  3.4  3.4  3.4  3.4  3.4  3.5  3.5  3.5
3.5  3.4  3.4  3.3  3.3  3.2  3.2  3.2      3.5  3.5  3.5
3.4  3.3  3.3  3.2  3.1            3.8  3.7  3.6  3.5  3.5
3.4  3.3  3.2              3.8  3.8  3.7  3.6  3.5  3.5  3.5
3.5              3.7  3.7  3.7  3.6  3.5  3.5  3.5  3.5  3.5
3.5  3.5  3.6  3.6  3.7  3.6  3.5  3.5  3.5  3.5  3.5  3.5
3.5  3.5  3.5  3.5  3.6  3.5  3.5  3.5  3.5  3.5  3.5  3.5
```

Figure 4.  In two dimensions it is possible for discontinuities not to partition data.

discontinuities (which do not partition the data), it is possible to use classification methods to identify and locate the discontinuities. The simplest approach is to use the methods for one-dimensional segmentation, on a series of parallel sections across the area of interest, in one direction or two perpendicular directions. If discontinuities are clear-cut, little noise is in the data, and if all discontinuities are at sufficiently high angles to the lines of section, it should be possible to trace them across the area. However, the method does not guarantee closed segments in two dimensions. To achieve two-dimensional segmentation, a more powerful technique must be used, such as one of the many cluster-analysis methods, to classify the data into a number or sets of contiguous points.

## CLUSTER ANALYSIS

The common aim of all cluster-analysis methods is to partition multivariate data into an unspecified number of sets, given a mimimum of a priori information. The procedures are well suited to problems such as biologic taxonomy or to the classification of rock samples.

A wide variety of clustering methods are assessed critically by Everitt (1974). Most of these methods yield unconstrained solutions, even if they use specific initial states of classification. However, for identification of discontinuities by zonation of spatially distributed data, such unsupervised clustering is not suitable. Clusters representing zones must consist of sets of contiguous points.

Openshaw (1974) published a clustering algorithm that incorporates a contiguity constraint for a gridded data set. Extension to irregularly distributed data is simple requiring only the construction of a contiguity matrix (the adjacency matrix of graph

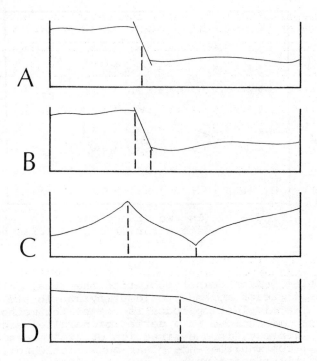

Figure 5. Types of discontinuity. *A* a fault considered as single discontinuity; *B*, fault considered as two discontinuities; *C*, ridges and valleys; *D*, breaks in slope.

theory) in conjunction with any of the clustering criteria. Thus it is possible to segment spatially distributed data comprising gridded or irregularly distributed data points, using a vector z of values at each data point (this obviously includes the situation of a single z value as discussed in another part of this paper).

Such an approach is powerful in concept, and allows a multivariate approach, but it must be remembered that it uses a restrictive definition of "discontinuity".

## MINIMAL-SPANNING TREE

Zahn (1971) suggested a clustering method based entirely on the concepts of graph theory in which a set of irregularly spaced variate data is linked to form a spanning tree. A graph is connected, so that every point is linked to some other point in the graph, and it contains no cycles, that is there is only one possible path along the edges (linkages) from any node to any other node. Each edge has an associated weight, and the minimal-spanning

Table 1. Types of discontinuity and techniques which may be used in their identification.

| Technique | *Classification* | *'Highlighting'* |
|---|---|---|
| Discontinuity: | | |
| Fault as a single discontinuity | Yes | Yes |
| Fault as two discontinuities | No* | Yes |
| Ridges and valleys | No* | Yes |
| Breaks in slope | No* | Yes |

* unless using a piecewise regression method.

tree is the spanning tree in which the sum of these weights is lowest. If the weights are defined as the Euclidean distances between points, the minimal-spanning tree will join the points which are closest together. By using a contiguity constraint such as that suggested by Openshaw (1974) and defining the weights as some function of the absolute difference, or the gradient in z value between two points, or as a generalized distance measure between the values of z vectors at two points, it is possible to construct a minimal-spanning tree of use in discontinuity identification. By successively deleting edges with the highest weight, the data points may be partitioned to obtain a zonation of the data. This devisive classification approach yields results similar to those of related cluster-analysis methods, and suffers from the same theoretical disadvantages.

## THE METHOD OF PAVLIDIS

Pavlidis (1972) also uses graph-theory concepts. First a one-dimensional piecewise linear-regression segmentation is applied to each of a set of parallel traverses across the area of interest. Each segment is labelled with the estimated slope, and a graph is constructed in which adjacent segments are joined if the difference between their slopes is smaller than a predefined threshold. Each connected subgraph then represents one segment or zone. A problem with this method is that the slope (regression coefficient) values refer only to the component of gradient in the direction of the parallel traverses.

## METHODS FOR HIGHLIGHTING DISCONTINUITIES

If the constraint that discontinuities must bound separate zones is relaxed, and the possibility of discontinuities which do not partition the data is allowed, it is necessary to adopt a nonclassification approach. A technique allows rapid subjective identification of discontinuities by graphic or computational enhancement of such features, should be used. Many techniques are available that may be used to achieve this end, including optical techniques (outside the scope of this paper), simple contouring, in which discontinuities are identified by looking at the spacing of contour lines, and more sophisticated methods of pattern recognition and image enhancement.

Contour plotting gives a moderately good graphic presentation of the simplest types of discontinuity - sudden changes in level, ridges, valleys - but the more subtle types, such as minor breaks in slope, are difficult to detect by examination of a contour map alone. Analytical hill-shading, as developed by Yoeli (1965), particularly with light incident vertically on the surface of interest, gives no z-value (altitude) data, but emphasizes any discontinuities present by differences in shading. A number of techniques, specifically designed to assist identification of discontinuities, may be derived from Yoeli's method.

Finally, some new pattern recognition techniques may be directly applied to the problem. McCormick and Jayaramurthy (1975) described an algorithm, based on statistical decision theory, for analyzing textures of digitized (gridded) data. Their method includes routines for border detection that are directly applicable to the problem of discontinuity analysis.

## ANALYTICAL HILL-SHADING

This method usually is applied to gridded data, although it is possible to use a hill-shading method directly on raw-data points by using an irregular triangular "grid". The method developed by Yoeli (1965) is based on Wiechel's (1878) theory of surface illumination. The light intensity at a given surface location is

$$I = \cos e$$

where $e$ is the angle between the incident light and the normal to the surface.

For discontinuity analysis, incident light may be used from any angle, but lacking any prior information, the best inclination is vertical. Thus the light intensity reduces to

$$I = \cos d$$

where $d$ is the angle of dip. The darkest grid cells then are those with the steepest slopes. It is possible to compute approximations to the first or second derivative of the surface at any point on the surface, and thus the first or second derivative surfaces can be shaded. The three alternative methods are illustrated in Figure 6. Choice of method depends solely on the type of discontinuity one is looking for: whether the steepest slopes, or the positions at which the slope changes most rapidly, or the positions at which the value of the first derivative (i.e. rate of change of slope) changes more rapidly.

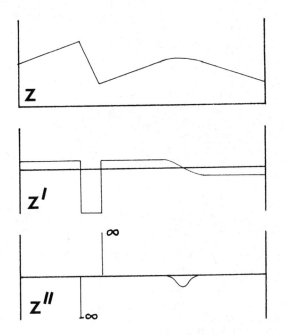

Figure 6. Fault (seen in z-value plot) appears as single blip in first derivative, z'; two sides of blip are identified separately in second derivative, z''. Smooth curve is reflected in first and second derivatives as smaller feature.

## SPECIFIC METHODS

A number of different methods fall in this category, but they have not been evaluated in detail. Up to three logical steps are involved in each of these techniques. First, a gridding algorithm may be required to generate either a regular rectangular grid or an irregular triangular grid; second, the data must be processed in some manner to emphasize any discontinuities which may be present; and third, some method is required to aid interpretation of the results from the second stage, either by a graphic display or by some objective identification technique.

(i) *Gridding*. A large number of interpolation algorithms have been proposed for contouring. Most of these smooth the data to some extent and thus tend to obscure the discontinuities. Moreover, some algorithms, such as the unweighted moving average, introduce spurious discontinuities which are not always noticed when contouring the data, but confound any attempt to discover genuine discontinuities. Thus, it is best to use interpolation techniques based on continuous functions, such as various splining methods, or weighted moving-average methods in which the weighting is a continuous function of distance from the grid point whose value is being estimated.

(ii) *Emphasis of discontinuities*. Analytical hill-shading, which has been discussed, is one such method for emphasizing discontinuities.

There is a group of methods based on comparison of adjacent grid cells (rectangular or triangular) using some criterion which correlates with any discontinuities in the data. The simplest approach is to approximate the first derivative of the data by computing the difference in absolute z value in adjacent cells; a second iteration will yield estimates of the second derivative. If the grid is irregular, gradients must be used rather than differences, to take into account the differing distances between grid points. An alternative approach is to consider the angle or radius of curvature as defined by adjacent grid cells.

Another group of methods is based on the concept of a local index of variability of the data. Each grid cell is assigned a value which is perhaps the standard deviation, coefficient of variations, or the range, or values at its corner points. This is a direct analog of moving-average methods, and so would be "moving standard deviations", etc. There is an obvious link here with the concepts of geostatistics, in which variance is dependent on spatial position, but the emphasis is different, because no assumptions of stationarity can be made. The variogram cannot be used because the suspected presence of discontinuities in the data invalidates any variogram which might be computed.

(iii) *Aids to interpretation*. Two approaches are possible: simply display the results from the second stage processing, or attempt an objective identification of any discontinuities present. To display the results, all the methods of computer graphics are available - contouring, gray-scale plotting (Howarth, 1971), perspective plots, etc. The peaks on these plots will represent the most probable areas of discontinuity in the data.

Objective identification of discontinuities is more difficult. Having obtained a grid of values which represent some index of discontinuity of the data, cluster analysis may be used (as discussed in the context of the raw data). The clustering is likely to be more informative if one uses some technique for peak intensification or image clarification on the grid before entering the clustering algorithm. A number of these methods have been developed for pattern-recognition studies and for improving the quality of digitized images. McCormick and Jayaramurthy (1975) illustrate the effectiveness of such a technique in the context of texture analysis. It seems that the best method of objectively extracting discontinuities would be to use an image-cleaning technique followed by cluster analysis with Openshaw's contiguity constraint, to obtain as separate clusters all the zones in which the "discontinuity index" has high values.

## TEXTURE ANALYSIS

An approach which may be of great value in the analysis of areally distributed geological data (not merely to assist the identification of discontinuities) was described by McCormick and Jayaramurthy (1975). They have applied decision-theoretic criteria to the problem of discriminating between different surface textures (as represented in the computer by digitized patterns).

The relevance of their technique to the identification of discontinuities lies in the power of the algorithm to detect the borders between zones of differing texture. The range of problems which can be analyzed by this technique is broader than is susceptible to the methods discussed here. It is possible not only to distinguish between zones of different z value, but also to separate, for example, rough and smooth topographic regions, or photogeological textures resulting from contrasting lithologies. As previously mentioned, it is possible to "clean" the picture which is generated, for instance, by removing points which do not have a certain number of neighbors of the same class. In this manner spurious discontinuities due to noise may be eliminated.

## CONCLUSIONS

Practical assessment of all the techniques described in a geological context suggests further possibilities. One line of inquiry which seems to be promising is in the relationship between discontinuity analysis and spatial autocorrelation techniques. Contiguity constraints derived from discontinuity identification techniques could be applied to data whose autocorrelation characteristics are to be studied. Many of the methods described in this paper are being implemented in generalized programs within the G-EXEC data-base management system (Jeffery and others, 1975) and a paper on the application of these methods is in preparation.

## ACKNOWLEDGMENTS

This paper is published with the permission of the Director, Institute of Geological Sciences. I wish to thank Dr. D.G. Farmer for suggesting some of the concepts incorporated in this paper, and Prof. D.F. Merriam and Dr. T.V. Loudon for their most helpful criticisms of the manuscript.

## REFERENCES

Everitt, B., 1974, Cluster analysis: Heinmann, London, 122 p.

Hawkins, D.M., 1972, On the choice of segments in piecewise approximation: Jour. Inst. Math. Applications, v. 9, no. 2, p. 250-256.

Hawkins, D.M., and Merriam, D.F. 1973, Optimal zonation of digitized sequence data: Jour. Math. Geology, v. 5, no. 4, p. 389-395.

Hawkins, D.M., and Merriam, D.F., 1974, Zonation of multivariate sequences of digitized geologic data: Jour. Math. Geology, v. 6, no. 3, p. 263-269.

Hawkins, D.M., and Merriam, D.F., 1975, Segmentation of discrete sequences of geologic data: Geol. Soc. America Mem. 142, p. 311-315.

Howarth, R.J., 1971, FORTRAN IV program for grey-level mapping of spatial data: Jour. Math. Geology, v. 3, no. 2, p. 95-121.

Jeffery, K.G., Gill, E.M., Henley, S., and Cubitt, J.M., 1975, The G-EXEC system users manual: Inst. Geol. Sciences and Atlas Computer Lab., Issue no. 2, 136 p.

Kruskal, J.B., 1964, Nonmetric multidimensional scaling: A numerical method: Psychometrika, v. 29, p. 115-129.

Kulinkovich, A.Ye., Sokhranov, N.N., and Churinova, I.M., 1966, Utilization of digital computers to distinguish boundaries of beds and identify sandstones from electric log data: Intern. Geology Rev., v. 8, no. 4, p. 416-420.

McCormick, B.H., and Jayaramurthy, S.N., 1975, A decision theory method for the analysis of texture: Intern. Jour. Computer and Information Sciences, v. 4, no. 1, p. 1-38.

Openshaw, S., 1974, A regionalisation program for large data sets: Computer App., v. 1, nos. 3-4, p. 136-160.

Pavlidis, T., 1972, Segmentation of pictures and maps through functional approximation: Computer Graphics and Image Processing, v. 1, p. 360-372.

Sammon, J.W., Jr., 1969, A nonlinear mapping for data structure analysis: IEEE Trans. Computers C-18, no. 5, p. 401-409.

Testerman, J.D., 1962, A statistical reservoir-zonation technique: Jour. Petroleum Tech. (Aug. 1962), p. 889-893.

Webster, R., 1973, Automatic soil-boundary location from transect data: Jour. Math. Geology, v. 5, no. 1, p. 27-37.

Wiechel, K., 1878, Theorie und Darstellung der Beleuchtung von nicht gesetzmaessig gebildeten Flaechen mit Ruecksicht auf die Bergzeichnung: Civilingenieur, v. 24, p. 335-364.

Yoeli, P., 1965, Analytical hill-shading: Surveying and mapping (Dec. 1965), p. 573-579.

Zahn, C.T., 1971, Graph-theoretical methods for detecting and describing gestalt clusters: IEEE Trans. Computers, v.C-20, no. 1, p. 68-86.

# THE DISPLAY OF THREE-FACTOR MODELS

W.E. Stephens

*University of St. Andrews*

ABSTRACT

A method is described in which the positions of factor loadings or factor scores may be seen in relation to three factor axes. The projection simulates perspective by using circles to represent points and by decreasing the radius of the circle with distance from a selected viewpoint anywhere outside the space occupied by the axes. KEY WORDS: *graphics, factor analysis, principal components analysis, geochemistry, sedimentology.*

INTRODUCTION

Interpretation of the factors generated by principal components analysis and factor analysis generally is aided by locating the variables or samples in the space defined by paired combinations of these factors. It may be the situation that problems involving multivariate data satisfactorily reduce to three factors and their relationships are examined by plotting all possible pairs of factors. In a planar representation of the three factors, the third dimension can be achieved by using perspective in the diagram. A perspective plot of this type is not suitable for quantitative examination but may have value for qualitative interpretation of the distribution of variables or samples through the factors. This paper describes the application of one type of projection to examples of both R- and Q-mode analyses.

MOLECULE REPRESENTATIONS

One method of representing a point in space is to consider it as the center of a sphere. The perspective element can be introduced by changing the size of the sphere in relation to the distance from the viewing point. This technique may be used in diagrammatic representations of molecules in which atoms are represented by spheres proportional in size to their atomic radii, whereas the spheres diminish in size with distance from the viewing position. Bonds are represented by lines tapered to heighten the effect of perspective. Computer programs exist which produce such diagrams on a digital graph plotter and the input can be adapted to plot three orthogonal axes in any viewing position. All samples or variables can be located at the centers of circles within this space. The program used in this study was PAMOLE (Cole and Adamson, 1969).

## EXAMPLES

Imbrie and Purdy (1962) classified modern Bahamian carbonate sediments in one of the earliest examples of the application of Q-mode factor analysis to a geological problem. Some 40 samples from five main facies were analyzed for their constituent particle compositions and four factors were chosen from the rotated quartimax factor matrix. Of the six possible factor pairs, four were represented in scatter diagrams. Although good separation of facies was achieved, no single two-factor plot gave complete resolution and there was some overlap of groups. Projection of the first three factors using the method described here, at a view point (10,40,10) for factor one, factor two, and factor three respectively, is given in Figure 1. It can be seen from this

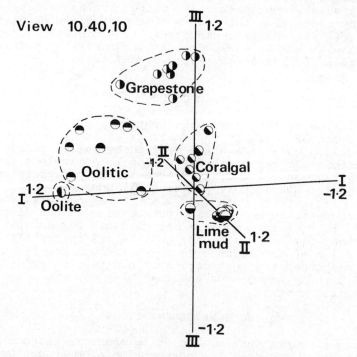

Figure 1. Plot of first three factors from Imbrie and Purdy (1962). View 10,40,10 indicates distance along each axis from center of I, II, and III respectively.

diagram that, providing a suitable viewing position is chosen, resolution of all samples into facies with no group overlap is possible on a single plot.

The relationships between 16 major and minor chemical elements in 33 samples from a Lower Paleozoic condensed marine shale succession have been examined by principal components analysis (Stephens and others, 1975). Three principal components account for almost 80 percent of the variance and Figure 2 shows the contributions of all elements to the factors. Clay, carbonate, metal, and silica associated elements are distinct and the diagram indicates the independent or complex geochemical behaviors of Y, Mg, and Na.

## VIEW

The viewpoint is specified in terms of distances along each of the three coordinates, and the view is taken along the line joining the viewpoint to the origin. It may be worthwhile generating plots from the eight viewing positions which are equidistant along all coordinates. Clusters of circles should be examined carefully

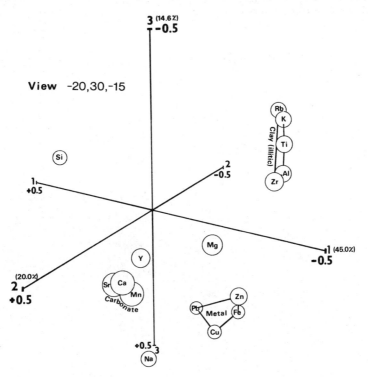

Figure 2. Plot of first three principal components from Stephens and others (1975). Note axes have been inverted. Distances are measured along axes in atomic units.

for their true spatial relationships. In Figure 2 for instance, Zn seems to be associated closely with the other metals Pb, Cu, and Fe. Its larger size, however, reflects in particular its lower positive loading for the third principal component. The orientation of the axes also has an influence on the ability of many observers to perceive the third dimension. It seems that geologists appreciate spatial relationships in this type of plot when they resemble conventional crystallographic diagrams.

REFERENCES

Cole, A.J., and Adamson, P.G., 1969, A simple method for drawing molecules using a digital plotter: Acta Crystallographica, v. A25, pt. 4, p. 535-539.

Imbrie, J., and Purdy, E.G., 1962, Classification of modern Bahamian carbonate sediments, *in* Classification of carbonate rocks: Am. Assoc. Petroleum Geologists Mem. 1, p. 253-272.

Stephens, W.E., Watson, S.W., Philip, P.R., and Weir, J.A., 1975, Element associations and distributions through a Lower Paleozoic graptolite shale sequence in the Southern Uplands of Scotland: Chemical Geology, v. 16, no. 4, p. 269-294.

# INDEX

abnormal fluid pressure, 105
"agglomerative-polythetic" strategy, 25
Allen's ideal alluvial cycle, 151
analysis of variance, 13
analytical hill-shading, 164
association analysis, 23
autocorrelation, 157

basin response, 106
Bouma's ideal turbidite sequence, 151
boundary, 110

Canadian Shield, 12
canonical variate, 88
carbonate petrology, 23
chi-square statistic, 25
classification, 23, 61, 74, 143, 157
cluster analysis, 11, 29, 37, 73, 121, 157, 161, 166
contiguity matrix, 161
cophenetic values, 39
correspondence analysis, 57
cross-association, 33

data
   display, 121
   independence, 2
   mapping, 157
   systems, 1
delta profile, 106, 108, 113
dendrogram, 19, 27, 38, 41, 45, 80, 81
Dice coefficients, 144
digitized images, 166
discontinuity analysis, 157
discriminant analysis, 73, 87, 121
"divisive-monothetic" strategy, 25
Dungeness, Kent, 123

Euclidean distance, 37, 40, 146

F-test statistic, 86, 88
factor analysis, 53, 61, 64, 67, 69, 169, 170
Fourier analysis, 121, 125
fuzzy-set analysis, 121, 129

G-EXEC, 2, 167
geochemical analyses, 56, 61, 169
geophysics, 99
gradient analysis, 82
grain-size data, 93
graph theory, 162
graphics, 53, 169
gridding, 161, 165
groundwater, 99
Gulf of Mexico, 112

history matching, 100
hydrology, 99

isostatic subsidence, 106
Israel, 23

Jaccard coefficient, 144

Kansas, 11
Kincardine Basin, 35, 44
Kruskal's "stress", 146

Lake Bonneville, 117
lakes, 61, 67
Limestone Coal Group, 35, 44
Long Island Sound, 74
Louisiana, 112

machine independence, 2
map comparison, 19
mapping, 121
marine sediments, 73
measurement scales, 143, 145
mechanical-size analysis, 78
meteorology, 99
mineralogy, 61
minimal-spanning tree, 162
mining, 99
Minnesota, 61
modular design, 3

# INDEX

molecule representations, 169
monotonic regression, 146
multidimensional scaling, 143, 149
multivariate mixing, 121
multivariate-mixture analysis, 136
multivariate statistics, 54, 74, 144, 158

Niger Delta, 115
nonlinear least squares, 124
nonlinear mapping, 33, 39, 128
nonmetric multidimensional scaling, 145
numerical analysis, 99, 143
numerical taxonomy, 37

oceanography 73, 99
ordination, 73, 74, 82, 149

paleoenvironments, 23, 151
pattern-recognition, 110, 126, 129
Pavlidis method, 163
petroleum, 99
petrology, 99
phi variables, 80
polynomial discriminant function, 136
porosity-depth curve, 118
Portugal, 53
presence-absence data, 23
principal components analysis, 73, 79, 169
principal coordinates, 56
process programs, 7
proximity measures, 148

random-number generator, 102
ranking, 144
redundancy analysis, 80
reservoir sandstones, 147
ROKDOC, 40

sample classification, 80
sandstones, 143, 144

scatterdiagram, 88
Scotland, 35
sedimentation environments, 151
sedimentary thickness data, 11
sedimentology, 61, 99, 105
segmentation of sequence data, 158
similarity measures, 144
simulation, 99, 105
size-component analysis, 122
size-frequency distribution, 124
software, 6
soil science, 99
Sorensen's coefficient, 79
sorting, 82
standardize, 64, 126
statistics, 121
stepwise discrimination, 88
stratigraphy, 53, 105
structural development, 11
structure, 11, 157
surfaces, 159
system
    concepts, 2
    design, 2

testing and verification, 84
textural maturity classification, 150
texture analysis, 167
transition probabilities, 33, 36, 40, 46
trend-surface analysis, 11, 12, 73, 86

U-statistic, 88

varimax rotation, 64

Wilks' lambda, 74, 84

Yates' correction, 26

Ziqlag Formation, 23